S0-BIF-469

ADVANCED ASTROPHYSICS

This book develops the basic underlying physics required for a fuller, richer understanding of the science of astrophysics and the important astronomical phenomena it describes. The Cosmos manifests phenomena in which physics can appear in its most extreme, and therefore more insightful, forms. A proper understanding of phenomena such as black holes, quasars and extrasolar planets requires that we understand the physics that underlies all of astrophysics. Consequently, developing astrophysical concepts from fundamental physics has the potential to achieve two goals: to derive a better understanding of astrophysical phenomena from first principles and to illuminate the physics from which the astrophysics is developed. To that end, astrophysical topics are grouped according to the relevant areas of physics. The book is ideal as a text for graduate students and as a reference for established researchers.

The author obtained his PhD in 1984 from the University of Toronto where he earned the Royal Astronomical Society of Canada Gold Medal for academic excellence. After a brief postdoctoral stint at the University of British Columbia, he joined the faculty at the University of New Mexico where he pursued his interests in radio astronomy. He has been teaching for the past 17 years, earning an "excellence in teaching" award for the graduate courses on which this book is based. Dr. Duric has over 100 scientific publications and has authored and/or edited five books. In addition, he has developed a number of online classes, including a completely interactive, web-based freshman astronomy course. He is the recipient of the "Regent's Fellowship", the highest honour that UNM bestows on its faculty. His research has taken him around the world to over a dozen countries, accounting in part, for the global perspective that characterizes his book. Dr. Duric is a member of the American Astronomical Society and the Canadian Astronomical Society.

ADVANCED ASTROPHYSICS

NEB DURIC

University of New Mexico

CAMBRIDGE
UNIVERSITY PRESS

PUBLISHED BY THE PRESS SYNDICATE OF THE UNIVERSITY OF CAMBRIDGE
The Pitt Building, Trumpington Street, Cambridge, United Kingdom

CAMBRIDGE UNIVERSITY PRESS
The Edinburgh Building, Cambridge CB2 2RU, UK
40 West 20th Street, New York, NY 10011–4211, USA
477 Williamstown Road, Port Melbourne, VIC 3207, Australia
Ruiz de Alarcón 13, 28014 Madrid, Spain
Dock House, The Waterfront, Cape Town 8001, South Africa

http://www.cambridge.org

© Neb Duric 2004

This book is in copyright. Subject to statutory exception
and to the provisions of relevant collective licensing agreements,
no reproduction of any part may take place without
the written permission of Cambridge University Press.

First published 2004

Printed in the United Kingdom at the University Press, Cambridge

Typeface Times 11/14 pt.   *System* LATEX 2$_\varepsilon$ [TB]

A catalog record for this book is available from the British Library

Library of Congress Cataloging in Publication data

Duric, Nebojsa.
Advanced astrophysics / Neb Duric.
p.   cm.
Includes bibliographical references and index.
ISBN 0 521 81967 9 – ISBN 0 521 52571 3 (paperback)
1. Astrophysics.   I. Title.
QB461.D87 2003
523.01–dc21   2003048457

ISBN 0 521 81967 9 hardback
ISBN 0 521 52571 3 paperback

The publisher has used its best endeavors to ensure that the URLs for external websites referred to in this book
are correct and active at the time of going to press. However, the publisher has no responsibility for the websites
and can make no guarantee that a site will remain live or that the content is or will remain appropriate.

Contents

Preface

Astrophysics strives to describe the Universe through the application of fundamental physics. The Cosmos manifests phenomena in which the physics can appear in its most extreme, and therefore more insightful, forms. Consequently, developing astrophysical concepts from fundamental physics has the potential to achieve two goals: to derive a better understanding of astrophysical phenomena from first principles, and to illuminate the physics from which the astrophysics is developed. To that end, astrophysical topics are grouped, in this book, according to the relevant areas of physics. For example, the derivation of the laws of orbital motion, used in the detection of extrasolar planets, takes place in the classical mechanics part of the book while the derivation of transition rates for the 21 cm neutral hydrogen line, used to probe galaxy kinematics, is performed in the quantum mechanics part. The book could serve as a text for graduate students and as a reference for established researchers.

The content of this book is based on the material used by the author in support of advanced astrophysics courses taught at the University of New Mexico. The intended audience consists of graduate students and senior undergraduates pursuing degrees in physics and/or astrophysics. Perhaps the most directly relevant demographic is the combined Physics and Astronomy departments. These departments tend to emphasize the fundamental physics regardless of the research track pursued by the student. In many cases a separate astrophysics degree is not an option. In these departments (such as the author's) all students must pass the same physics comprehensive examination. Consequently, students must be well prepared in fundamental physics both from the points of view of course work as well as research. In the latter case a strong physics foundation is very helpful in developing thesis topics to an acceptable level in a physics-dominated department. This book is specifically aimed at those departments.

The department of Physics and Astronomy at the University of New Mexico requires its graduate students to take the physics comprehensive exam. Courses based on the material in this book have helped astrophysics and physics students prepare for these exams. I attribute this benefit to the fact that the astrophysical topics provide interesting and insightful manifestations of the fundamental physics, of which, the students previously may have had only a theoretical knowledge. I therefore expect the book to impact the physics as well as the astrophysics students in the mixed departments. I also expect that graduate students in physics-only and astronomy-only departments may choose to use this book to hone their research skills. Targeting junior/senior undergraduates is also possible in schools where the science curricula are robust.

Multi-disciplinary and cross-disciplinary investigations are playing an increasingly important role in scientific research. The cross-over of particle physicists into cosmology, and the establishment of the field of astroparticle physics, is just one manifestation of the growing overlap between physics and astrophysics disciplines. The emphasis on the linkage of fundamental physics and astrophysics makes this book potentially useful as a reference to physics and astrophysics researchers who wish to broaden their research base.

The author acknowledges the help of Dr. Rich Epstein (Los Alamos National Laboratory) who co-taught the first course in the series of courses that have led to the development of the material used in this book. The cooperation and help of the many students who have taken these courses has been instrumental in identifying many typos and inconsistencies in the course material. Finally, the author acknowledges the help and support of the department of physics and astronomy at UNM and the patience and encouragement of family and friends in this endeavor.

Part I

Classical mechanics

The visible Universe contains hundreds of billions of galaxies, each consisting of billions of stars. Recent discoveries of extrasolar planets lead us to believe that a typical galaxy may contain billions of planets (and presumably, asteroids and comets). The planets, stars and galaxies interact on a hierarchy of scales ranging from AU to parsec to megaparsec, experiencing forces arising from gravity, on all scales, and cosmic expansion on the larger scales. The combination of gravitational attraction and cosmic expansion has shaped the visible matter in the Universe into a hierarchy of structures leading to clusters and superclusters of galaxies.

A full description of the interactions that define the large-scale structure of the Universe and its constituent parts requires the application of general relativity on all scales and the introduction of a new force, as embodied in the recently proposed cosmological constant, on the largest scales. In this part, however, we limit ourselves largely to the application of classical (Newtonian) mechanics which is sufficiently accurate to describe the topics covered in this part and has the advantage of being more intuitive and accessible to the reader.

This part begins with a review of the basic elements of classical mechanics, subsequently used to derive Kepler's laws, the Virial theorem and various aspects of orbital motion. The resulting derivations are applied to specific astrophysical problems such as planetary motion, extrasolar planets, binary stars, galaxy rotation curves, dark matter, the large scale structure of the Universe and cosmic expansion.

Chapter 1

Orbital mechanics

I begin this part by reviewing some basic concepts that underlie Newtonian gravitation. The concepts of universal gravitation, center of mass and reduced mass are defined and subsequently used in the following chapters.

1.1 Universal gravitation

The gravitational force acting between two bodies, m_1 and m_2, located at $\vec{R}_1$ and $\vec{R}_2$, is given by

$$\vec{F} = \pm \frac{Gm_1m_2}{|\vec{R}_1 - \vec{R}_2|^3}(\vec{R}_1 - \vec{R}_2) \tag{1}$$

where the quantities are defined in Fig. 1.1 and G is the gravitational constant. The $\pm$ signs reflect the fact that the same magnitude of force acts on m_1 and m_2 but with opposite sign.

1.1.1 Center of mass

Consider a point on a line, joining m_1 and m_2, which is the centroid of the total mass distribution. We call this centroid the *center of mass* of the two-body system. The vector $\vec{r}$, separating the two masses, can then be decomposed into $\vec{r}_1$ and $\vec{r}_2$ relative to the center of mass, such that

$$\vec{r} = \vec{r}_1 - \vec{r}_2.$$

From Newton's Second Law

$$\vec{F}_1 = m_1\ddot{\vec{r}}_1 = -\frac{Gm_1m_2}{|\vec{r}|^3}\vec{r}. \tag{2}$$

3

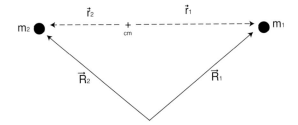

Fig. 1.1 The universal law of gravitation. Gravity is a mutual force that acts between the masses m_1 and m_2.

Similarly

$$\vec{F}_2 = \frac{Gm_1m_2}{|\vec{r}|^3}\vec{r} \tag{3}$$

$$\Rightarrow \ddot{\vec{r}}_1 - \ddot{\vec{r}}_2 = \ddot{\vec{r}} = -\frac{G(m_1 + m_2)}{|\vec{r}|^3}\vec{r} = -\frac{GM}{|\vec{r}|^3}\vec{r}. \tag{4}$$

The acceleration of the two bodies toward each other is proportional to the total mass and inversely proportional to the square of the distance between them. The location of the center of mass (CM) can now be found

$$m_1\ddot{\vec{r}}_1 = -m_2\ddot{\vec{r}}_2 \Rightarrow -m_1\frac{GM}{|\vec{r}|^3}\vec{r}_1 = m_2\frac{GM}{|\vec{r}|^3}\vec{r}_2 \Rightarrow \vec{r}_1 = -\frac{m_2}{m_1}\vec{r}_2 \tag{5}$$

where r_1 and r_2 represent the distance of m_1 and m_2 from the center of mass, respectively. The center of mass is a useful concept in astronomy. It marks the center about which two astronomical bodies orbit. In an isolated two-body system, the center of mass is not seen to accelerate.

1.1.2 Reduced mass

Let us define a mass such that

$$\vec{F} = \mu\ddot{\vec{r}} = -\frac{GM\mu}{|\vec{r}|^3}\vec{r} = -\frac{Gm_1m_2}{|\vec{r}|^3}\vec{r}$$

$$\Rightarrow \mu = \frac{m_1m_2}{m_1 + m_2}. \tag{6}$$

The concept of reduced mass allows us to transform any two-body problem into a one-body problem where the reduced mass responds to a central force emanating from a point whose distance is equal to the separation of the original two bodies.

1.2 Kepler's laws

We are now in a position to derive the most famous orbital laws used in astronomy, Kepler's laws. We begin, as with so many other problems in classical mechanics, with the Lagrangian

$$L = T - V \tag{7}$$

where T is the kinetic energy and V is the potential energy. Let us set $m = m_1$ and $M = m_2$ in anticipation of defining planetary orbits where the planets have much lower masses than the Sun (that is $m \ll M$). We are considering a two-body interaction so that the expected motion is in a plane and possibly periodic. It therefore makes sense to use polar coordinates, r and θ for this problem. Equation (7) then becomes

$$L = \frac{1}{2}m(\dot{r}^2 + r^2\dot{\theta}^2) - V(r). \tag{8}$$

We are now in a position to determine the angular momentum p_θ from the Lagrangian. Recall that

$$p_\theta = \frac{\partial L}{\partial \dot{\theta}} = mr^2\dot{\theta}.$$

We now use the Lagrange equation of motion

$$\frac{\mathrm{d}}{\mathrm{d}t}\frac{\partial L}{\partial \dot{\theta}} - \frac{\partial L}{\partial \theta} = 0.$$

so that

$$\frac{\mathrm{d}}{\mathrm{d}t}(mr^2\dot{\theta}) = 0.$$

Integrating

$$mr^2\dot{\theta} = l = \text{constant}. \tag{9}$$

Equation (9) represents the *conservation of angular momentum*. Rearranging terms

$$\frac{1}{2}r^2\dot{\theta} = \frac{1}{2}\frac{l}{m} = \text{constant}.$$

Recall that the area of an elemental triangle is given by $\mathrm{d}A = r^2/2 \, \mathrm{d}\theta$, so that

$$\frac{\mathrm{d}A}{\mathrm{d}t} = \frac{r^2}{2}\left(\frac{\mathrm{d}\theta}{\mathrm{d}t}\right) = \text{constant}. \tag{10}$$

According to (10), a radius vector sweeps out equal areas in equal time which, of course, is *Kepler's Second Law*.

The Hamiltonian or total energy of a two-body system is given by

$$E = T + V = \frac{1}{2}m(\dot{r}^2 + r^2\dot{\theta}^2) + V(r). \tag{11}$$

Rearranging (11) and solving for $\dot{r}$

$$\dot{r}^2 = \frac{2}{m}(E - V(r)) - r^2\dot{\theta}^2.$$

But, $r^2\dot{\theta} = l/m$ so that

$$\dot{r}^2 = \frac{2}{m}(E - V(r)) - \left(\frac{l}{rm}\right)^2$$

$$\Rightarrow \dot{r} = \sqrt{\frac{2}{m}\left(E - V(r) - \frac{l^2}{2mr^2}\right)}.$$

The above can be solved for dt so that

$$dt = \frac{dr}{\sqrt{(2/m)(E - V(r) - (l^2/(2mr^2)))}}. \tag{12}$$

Equation (12) can now be used to determine the shape of the orbit resulting from the two-body interaction. What we really want is a function $r(\theta)$ which means converting (12) into a relationship between r and θ and eliminating t in the process.

We begin by noting that $r^2\dot{\theta} = r^2(d\theta/dt) = l/m$ so that $l\,dt = mr^2\,d\theta$

$$\Rightarrow \frac{d}{dt} = \frac{l}{mr^2}\frac{d}{d\theta}$$

so that

$$\frac{mr^2}{l}d\theta = \frac{dr}{\sqrt{(2/m)(E - V(r) - (l^2/(2mr^2)))}}$$

$$\Rightarrow d\theta = \frac{l\,dr}{mr^2\sqrt{(2/m)(E - V(r) - (l^2/(2mr^2)))}}$$

$$\Rightarrow \theta = \int_{r_0}^{r} \frac{dr}{r^2\sqrt{(2mE/l^2) - (2mV/l^2) - (1/r^2)}} + \theta_0. \tag{13}$$

Let $\mu = 1/r$ and substitute into (13)

$$\Rightarrow \theta = \theta_0 - \int_{\mu_0}^{\mu} \frac{d\mu}{\sqrt{(2mE/l^2) - (2mV/l^2) - \mu^2}}.$$

For $V = -(GmM)/r = -k/r = -k\mu$

$$\Rightarrow \theta = \theta_0 - \int_{\mu_0}^{\mu} \frac{d\mu}{\sqrt{(2mE/l^2) + (2mk\mu/l^2) - \mu^2}} \tag{14}$$

which can be put into standard form and solution with $\mu = x$

$$\int \frac{dx}{\sqrt{a + bx + cx^2}} = \frac{1}{\sqrt{-c}} \cos^{-1}\left[\frac{-b + 2cx}{q}\right] \tag{15}$$

where

$$q = b^2 - 4ac.$$

Comparison of (14) and (15) yields

$$a = \frac{2mE}{l^2} \qquad b = \frac{2mk}{l^2} \qquad c = -1$$

$$q = \left(\frac{2mk}{l^2}\right)^2 \left(1 + \frac{2El^2}{mk^2}\right)$$

so that the solution to (14) becomes

$$\theta = \theta' - \cos^{-1}\left[\frac{(l^2\mu/mk) - 1}{\sqrt{1 + (2El^2/mk^2)}}\right] \tag{16}$$

where θ' incorporates the additional constants resulting from the integration. Putting $\mu = 1/r$ back into (16) and taking the cosine of both sides yields

$$\frac{1}{r} = \frac{mk}{l^2}\left(1 + \sqrt{1 + \frac{2El^2}{mk^2}} \cos(\theta - \theta')\right). \tag{17}$$

We now have a solution, $r(\theta)$, that determines the shape of the orbit and clearly depends on the energy, E, and the angular momentum, l. This equation can be compared with the general expression for a conic section

$$\frac{1}{r} = C(1 + \epsilon \cos(\theta - \theta')). \tag{18}$$

By equating (17) to (18) we see that

$$C = \frac{mk}{l^2} \tag{19}$$

$$\epsilon = \sqrt{1 + \frac{2El^2}{mk^2}}.$$

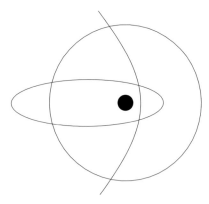

Fig. 1.2 Orbits as conic sections. Circular, elliptical and parabolic/hyperbolic classes of orbits are shown. The total energy, E, determines the class of orbit while the combination of E and the angular momentum, l, determines the shape of the orbit within a class. The Sun is shown as the small filled circle at the center.

The only variable that can be negative is the total energy of the two-body system so that

$$E > 0 \rightarrow \epsilon > 1 \quad \text{hyperbola}$$
$$E = 0 \rightarrow \epsilon = 1 \quad \text{parabola}$$
$$E < 0 \rightarrow \epsilon < 1 \quad \text{ellipse}$$
$$E = -\frac{1}{2}V = -\frac{mk^2}{2l^2} \rightarrow \epsilon = 0 \quad \text{circle.}$$

These define conic sections, as illustrated in Fig. 1.2.

In the solar system, planets have closed orbits ($E < 0$) and move in elliptical trajectories (*Kepler's First Law*). Kepler's Third Law can now be derived, beginning with the second law. Integrating (10) over a complete period of the orbit yields

$$\int_0^P \dot{A}\, dt = \frac{1}{2}\frac{l}{m}P = \pi ab \tag{20}$$

where πab is the area of an ellipse and a and b are the semi-major and semi-minor axes of the elliptical orbit. Now from (18) we can define a as the sum of distances that correspond to $\theta = \theta'$ and $\theta = \theta' + \pi$

$$a = \frac{1}{C(1 - \epsilon^2)}.$$

Combining this with the well-known relationship between a and b

$$b = a\sqrt{(1 - \epsilon^2)}$$

yields

$$b = \sqrt{\frac{a}{C}}. \tag{21}$$

Combining (19) and (21) yields

$$b = \sqrt{a}\sqrt{\frac{l^2}{mk}}. \tag{22}$$

Combining (20) and (22)

$$\frac{1}{2}\frac{l}{m}P = \pi a^{3/2}\sqrt{\frac{l^2}{mk}}$$

$$\Rightarrow P = 2\pi a^{3/2}\sqrt{\frac{m}{k}}$$

$$\Rightarrow P = \frac{2\pi}{\sqrt{GM}}a^{3/2}. \tag{23}$$

Equation (23) represents *Kepler's Third Law* – the square of the period is proportional to the cube of the diameter of the orbit.

1.2.1 Planetary orbits

The planets follow orbits as described by (18). However, the orbits differ significantly from each other and do not fall in exactly the same plane. Consequently, it is necessary to describe planetary orbits in three dimensions relative to a standard reference frame, as shown in Fig. 1.3.

There are two major reference points for a planetary orbit and both are related to the Earth. The Earth's orbit (plane NB) is used as the standard reference plane called the *ecliptic*. The intersection of the Earth's celestial equator with the ecliptic defines the *vernal* and *autumnal equinoxes*. The former is denoted as γ in Fig. 1.3. It is used as the fundamental reference point for defining the orbital elements. The plane of the Earth's orbit (the ecliptic) is $\gamma N'B$ while the plane of the planet's orbit is NQN'. The intersection of the two planes is called the *line of nodes* which connect the *ascending* and *descending nodes* (N and N', respectively – the direction of motion of the planet is indicated by the arrow). The Sun is located at the center and its position is denoted by S. The true orbit of the planet is shown as the ellipse pLA. The *perihelion* position is marked as A and the position of the planet, at time t, is denoted as p. The planet and the Sun define a radius vector, Sp, that cuts the great circle, NQN', at P1.

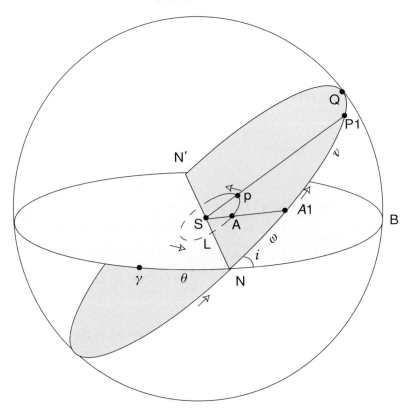

Fig. 1.3 The orbit of a planet relative to the Earth's orbit. A planetary orbit can be uniquely defined in 3-D space relative to γ and the Earth's orbit. The various parameters that characterize the planetary orbit are defined in the text.

With the help of Fig. 1.3, we can define the following parameters of the *apparent* orbit of the planet

$$v = A1 - P1 = \text{true anomaly}$$

$$\omega = N - A1 = \text{argument of perihelion}$$

$$\theta = \gamma - N = \text{longitude of ascending node}$$

$$\bar{\omega} = \theta + \omega = \text{longitude of the perihelion}$$

$$L = \theta + \omega + v = \text{true longitude of planet}$$

$$i = B - N - A1 = \text{inclination of orbit}$$

$$\tau = \text{time when planet is at perihelion, A.}$$

The six elements that completely define the orbit are $a, e, \theta, \bar{\omega}, i, \tau$. To complete the connection to (18), which we derived earlier, we see that $v = \theta - \theta'$. The

Table 1.1. *Planetary orbits – elements on January 1, 2000*

Planet	a (AU)	e	P (years)	i (degree)	θ (degree)	$\theta + \omega$ (degree)
Mercury	0.387	0.206	0.241	7.00	48.33	77.46
Venus	0.723	0.007	0.615	3.39	76.68	131.53
Earth	1.000	0.017	1.000	0.0	−11.26	102.95
Mars	1.524	0.093	1.85	1.85	49.58	336.04
Jupiter	5.203	0.048	11.862	1.31	100.56	14.75
Saturn	9.537	0.054	29.458	2.49	113.72	92.43
Uranus	19.191	0.047	84.012	0.77	74.23	170.96
Neptune	30.069	0.009	164.796	1.77	131.72	44.97
Pluto	39.482	0.249	246.378	17.14	110.30	224.07

additional elements allow us to determine the orbit relative to our perspective at the Earth. Table 1.1 lists the orbital elements of the planets in our solar system.

1.3 Binary stars

1.3.1 Visual binaries

Roughly half of all stars in the Galaxy are binaries. Analysis of binary star orbits via the equations we have derived thus far, provides valuable information regarding stellar properties and stellar evolution, information that would otherwise be difficult to obtain. Binary systems in which both stars are visible are known as visual binaries.

1.3.2 The apparent orbit

Binary stars represent the most general two-body problem. Their orbits are oriented randomly in space and are described fully in three dimensions in much the same way as were the planets we discussed earlier. However, because we only see a projection of the orbit on the sky we must somehow recover the orbital elements from an analysis of the 2-D orbit. The 2-D orbit is measured according to Fig. 1.4.

The most general form of an ellipse is given by

$$ax^2 + 2hxy + by^2 + 2gx + 2fy + 1 = 0 \tag{24}$$

where $x = \rho \cos\theta$ and $y = \rho \sin\theta$ and all coefficients are real constants. The equation of the apparent orbit is obtained by fitting (24) to a large number of measurements of ρ and θ. The more observations the better the fit and the more accurate the coefficients that define the shape of the apparent orbit. The procedures for

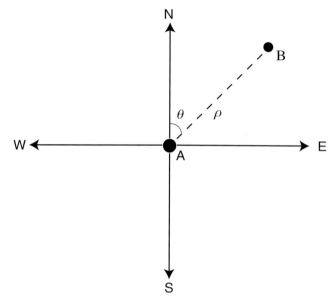

Fig. 1.4 Definition of the apparent orbit. The cardinal directions are shown. The apparent separation of the two stars is given by ρ and the position angle (measured east of north) is given by θ.

recovering the elements of the 3-D orbit from the measured 2-D orbit and the associated limitations are now discussed.

1.3.3 The true orbit

Consider Fig. 1.5, which shows the binary orbit relative to the observer. The position of the primary star of the binary system is denoted as S, at the center of the sphere. The position of the companion, at time t, is given by F. *Periastron* is at P. The plane LGM represents the plane of the binary orbit. The plane NLD corresponds to the *apparent* orbit. It is perpendicular to the line of sight to the observer and represents the projection of the plane of the true orbit in the direction of the observer. The line segment, SN, represents a reference angle, $\theta = 0$ (relative to true north for the observer). ML is the line of nodes defined by the intersection of the plane of the binary orbit and the plane NLD.

Analysis of Fig. 1.5 shows that

$$\rho = r \cos \mathrm{GD} \qquad (25)$$

where $r = \mathrm{SF}$ and ρ is the apparent distance of the companion from the primary (Fig. 1.4) and r is the true separation of the two stars. We also note the following

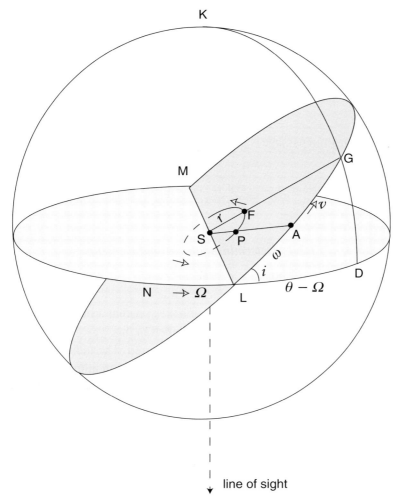

Fig. 1.5 Binary star orbit relative to the observer. The orbit of the secondary star is shown relative to the primary (S). The orbit is defined relative to a plane that is orthogonal to the line of sight to the observer.

relationship between three of the angles

$$\cos GD = \frac{\cos(v + \omega)}{\cos(\theta - \Omega)}. \tag{26}$$

Combining (25) and (26) yields

$$\rho = r \frac{\cos(v + \omega)}{\cos(\theta - \Omega)}. \tag{27}$$

Similarly,

$$\tan(\theta - \Omega) = \tan(v + \omega) \cos i. \tag{28}$$

The average angular velocity (averaged over one orbit) is given by

$$n = \frac{2\pi}{P}.$$

The time of periastron passage is given by τ. Kepler's Third Law must be used in its full form so that

$$n^2 a^3 = \frac{4\pi^2 a^3}{P^2} = G(m_1 + m_2). \tag{29}$$

1.3.4 Determining the orbital elements

The elements of the true orbit are a, e, i, Ω, ω, τ and P. The true orbit is determined by varying the orbital elements to generate values of ρ and θ using equations such as (27), (28) and (29) and comparing this with the apparent orbit until there is a good match. All the orbital parameters can be obtained in this way, including a **but** not a_1 and a_2 because the center of mass is not measured. The individual semi-major axes can only be obtained if the motions of the stars are monitored with respect to an absolute reference frame. Although this is possible the resulting elements tend to be much less accurate.

Determining the stellar masses

Since the individual semi-major axes are not measured, only the sum of the masses can be determined via (29). Again, if measurements are made on an absolute reference grid the individual masses can be obtained.

1.3.5 Spectroscopic binaries

When stars are too close to be individually resolved we rely on information obtained from their spectra. Generally one of the two stars dominates the observed spectrum by being the brighter of the two. We will therefore begin by considering a single spectrum.

The essential point is that the wavelengths of the spectral lines are measured repeatedly and accurately in order to search for Doppler induced wavelength variations. Consider Fig. 1.6, where G is the center of gravity of the two-body system. From the figure

$$z = r \sin(\text{PM}) \tag{30}$$

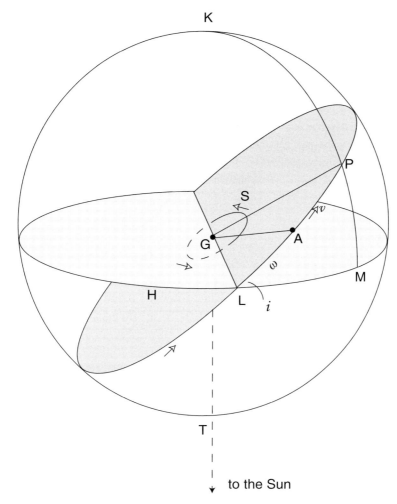

Fig. 1.6 Orbit of a spectroscopic binary relative to the observer. The orbit is defined relative to a plane orthogonal to the line of sight to the observer.

represents the elevation of the star above the HLM plane and therefore represents the component of r along the line of sight to the observer. It is only the line-of-sight component that contributes to the Doppler effect. Also, remember that the observer can be considered to be at infinity relative to the scale of the binary system.

Again, from Fig. 1.6

$$\sin(\text{PM}) = \sin(v + \omega) \sin i. \tag{31}$$

Combining (30) and (31)

$$z = r \sin(v + \omega) \sin i. \tag{32}$$

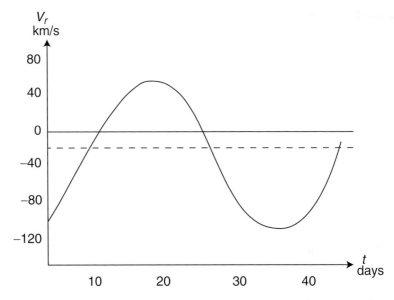

Fig. 1.7 Radial velocity curve showing the radial velocity changes as well as the systemic velocity. The systemic velocity is shown as a dashed line.

The *radial velocity* (the line-of-sight velocity) is simply the rate at which z changes with time plus whatever net velocity the binary system has with respect to the observer

$$V_r = V_s + \frac{dz}{dt}. \tag{33}$$

The velocity V_s is called the *systemic velocity* and represents the steady, non-periodic portion of the binary's motion along the line of sight. The velocity V_r is measured empirically from high-dispersion spectra. The velocity V_s is separated from dz/dt by examining *velocity curves* such as that shown in Fig. 1.7.

To extract useful information about the orbit of the star it is necessary to relate dz/dt to the parameters of the true orbit.

We begin with the general expression for an orbit (equations (18) and (20))

$$r = \frac{a(1 - e^2)}{1 + e \cos v} \tag{34}$$

and Kepler's Second Law

$$r^2 \frac{dv}{dt} = h = [n^2 a^4 (1 - e^2)]^{1/2}. \tag{35}$$

Differentiating both sides of (32) with respect to time yields

$$\frac{dz}{dt} = \frac{dr}{dt} \sin(v + \omega) \sin i + r \cos(v + \omega) \sin i \frac{dv}{dt}. \tag{36}$$

Doing the same thing with (34)

$$\frac{dr}{dt} = \frac{nae \sin v}{(1 - e^2)^{1/2}}. \tag{37}$$

From (35)

$$r\frac{dv}{dt} = \frac{1}{r}r^2\frac{dv}{dt} = \frac{h}{r} = \frac{na}{(1 - e^2)^{1/2}}(1 + e \cos v). \tag{38}$$

Substituting (37) and (38) into (36) yields

$$\frac{dz}{dt} = \frac{na \sin i}{(1 - e^2)^{1/2}} [\cos(v + \omega) + e \cos \omega] \tag{39}$$

which gives us what we have been looking for, the radial velocity as a function of the true orbital parameters. With many repeated measurements of dz/dt it is possible to determine the orbital parameters via (39). In practice, we vary the parameters in (39) in order to fit the observed radial velocity curve, such as the one shown.

One important aspect of this procedure is that it does not yield a unique value for i, the orbital inclination, because only the radial component of the velocity is measured without accompanying geometrical information. This means that only $a \sin i$ can be measured, not a. As we will see this affects the ability to measure the masses of the stars.

1.3.6 The mass function

From page 3 and the discussion of the center of gravity (center of mass), we have

$$\frac{a_1}{a_1 + a_2} = \frac{a_1}{a_0} = \frac{m_2}{m_1 + m_2}$$

so that

$$(a_1 \sin i)^3 = \left(\frac{a_0 m_2}{m_1 + m_2} \sin i\right)^3 = \frac{m_2^3 a_0^3 \sin^3 i}{(m_1 + m_2)^3}. \tag{40}$$

Re-examining Kepler's Third Law,

$$m_1 + m_2 = \frac{4\pi^2}{G} \frac{a_0^3}{P^2}. \tag{41}$$

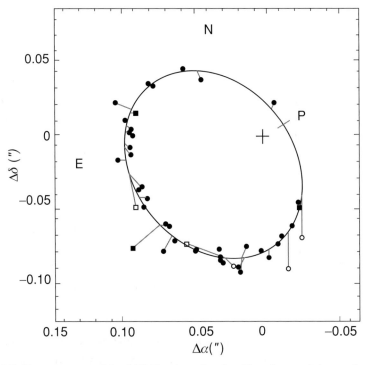

Fig. 1.8 The apparent orbit of 70 Tauri on the sky. The observed data points are shown along with the best-fit ellipse (from Torres, G., Stefanik, R.P. and Latham, D. W., 1997, ApJ, 479, pp. 268–278).

Combining (40) and (41), yields

$$\frac{m_2^3 \sin^3 i}{(m_1 + m_2)^2} = \frac{4\pi^2}{G} \frac{(a_1 \sin i)^3}{P^2}. \tag{42}$$

The quantity on the right can be estimated from observations. The left-hand side is referred to as the *mass function*. Though it contains information about the masses it is not possible to determine individual masses or, for that matter, the total mass. If two spectra are visible the ratio of the masses can also be obtained. If the binary is eclipsing ($i = 90°$) the individual masses can also be obtained.

1.3.7 Summary of binary star studies

Examples are now shown of results relating to binary stars. Figure 1.8 shows the orbit of 70 Tauri. The orbit was derived from speckle imaging, a method that compensates for the blurring of the Earth's atmosphere. The figure is taken from Torres *et al.* (1997). The system is also a double-line spectroscopic binary as shown in Fig. 1.9 (from Torres *et al.*, 1997). In all subsequent figures in this section, the dots represent the data and the curve represents the best fit model.

Table 1.2. *Binary orbits – derivable parameters*

Type of binary	Derivable parameters	Observations needed
Visual	Luminosity	Parallax, apparent brightness
	Sum of masses, orbital elements	Parallax, separation, period
	Individual masses	Absolute reference grid
Spectroscopic	Mass function, some orbital data	Single velocity curve
	Ratio of masses, some orbital data	Double velocity curve
Eclipsing	Radii, masses, orbital elements	Light and velocity curves

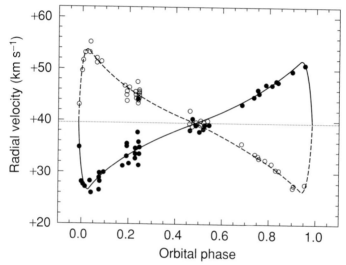

Fig. 1.9 Radial velocity curve of the double-line spectroscopic binary 70 Tauri. Observations of both stars are shown by the filled and unfilled circles. The systemic velocity is shown as a horizontal line (from Torres, G., Stefanik, R. P. and Latham, D. W., 1997, ApJ, 479, pp. 268–278).

For comparison, a single-line spectroscopic binary, 51 Tauri, is shown in Fig. 1.10, also from Torres *et al.* (1997). Table 1.2 summarizes and compares the various studies of binary stars.

1.3.8 Mass–luminosity relation

As noted earlier, one of the main reasons for studying binary stars is to determine the masses of stars and to correlate those masses with other properties. Such correlations yield important clues on how stars are born and how they evolve with time. A cornerstone for such studies is the mass–luminosity relation which is a correlation between the mass of a star and the rate at which it emits (and therefore produces) energy.

Orbital mechanics

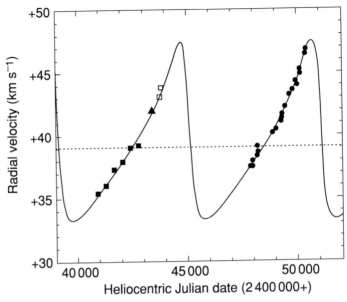

Fig. 1.10 Radial velocity curve of the single-line spectroscopic binary 51 Tauri. In this case only one star is visible (from Torres, G., Stefanik, R. P. and Latham, D. W., 1997, ApJ, 485, pp. 167–181).

The mass–luminosity relation is fairly tight and is well represented by the following equations

$$\frac{L}{L_\odot} = \left(\frac{M}{M_\odot} \right)^{4.0} \qquad (M > 0.43\ M_\odot)$$

$$\frac{L}{L_\odot} = 0.23 \left(\frac{M}{M_\odot} \right)^{2.3} \qquad (M < 0.43\ M_\odot).$$

An example of a mass–luminosity relation for members of the Hyades cluster of stars is shown in Fig. 1.11. It has been determined from mass measurements obtained using astrometric and spectroscopic techniques as described before. Figure 1.11 is also taken from Torres *et al.* (1997).

1.4 Extrasolar planets

The detection of extrasolar planets represents a holy grail of modern astronomy. The techniques used to hunt down planets are extensions of the methods used to study binary stars. I describe each of them briefly.

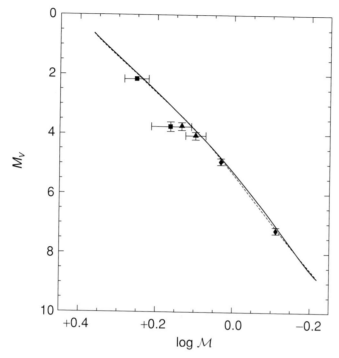

Fig. 1.11 The mass–luminosity relation for members of the Hyades star cluster. The visual magnitude is shown along the vertical axis while the log of the mass is shown along the horizontal axis (from Torres, G., Stefanik, R. P. and Latham, D. W., 1997, ApJ, 485, pp. 167–181).

1.4.1 The astrometric method

This method is essentially the same as the visual binary method except for the fact that the companion is not visible. Consequently, all measurements of the star must be made with respect to the center of mass, in other words in an inertial reference such as that provided by background stars. Figure 1.12 shows how such measurements might be made.

The measurements of the stellar position relative to the center of mass are, of course, angles (θ) and are related to the mass of the system as

$$\theta'' = \frac{m}{M} \frac{a(\text{AU})}{D(\text{pc})}$$

where θ'' is the angular deviation in seconds of arc, m is the mass of the planet, M is the mass of the star, a is the orbital semi-major axis of the planet relative to the center of mass of the system, in AUs, and D is the distance to the system in parsecs. This equation follows directly from (5) by setting $r_1 = \theta'' D$ and $r_2 = a$.

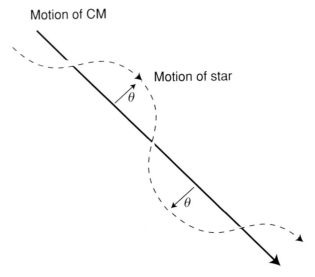

Fig. 1.12 Apparent motion of a star interacting with an unseen companion. The combination of the proper motion and the orbital motion of the star leads to the dashed path. The angle θ is the maximum deviation from a straight line.

The measured amplitude of the signal is independent of the inclination of the orbit which is the primary advantage of this method. The major disadvantages are that it is difficult to measure accurate positions in the presence of atmospheric turbulence and that it takes a long time to monitor changes in the orbit and accumulate a significant signal.

So far, the best candidate for this type of work is Lalonde 21185. George Gatewood (1996), using the Allegheny Observatory, has monitored this star for many years. He claims that the star has two Jovian-mass companions, one with an orbit of 30 years, the other 6 years. This result, however, is controversial. For details see the catalog of extrasolar planets, www.obspm.fr/encycl/catalog.html.

1.4.2   The radial velocity method

This method is identical to that used to study single-line spectroscopic binaries. It has proven to be the most prolific method for detecting extrasolar planet candidates.

From (42), assuming $m \ll M$ (which is true for planets)

$$\frac{m^3 \sin^3 i}{M^2} = \frac{n^2}{G} (a_1 \sin i)^3$$

but from (39), setting $a = a_1$ and $e = 0$, we have

$$V_r(\text{max}) = na_1 \sin i.$$

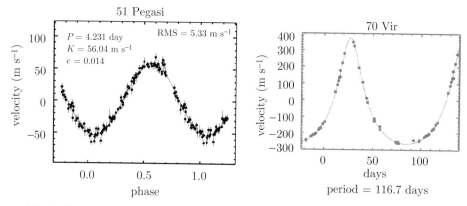

Fig. 1.13 The two stars, 51 Peg and 70 Vir, are responding to an unseen companion, as shown in these radial velocity plots. These curves are equivalent to the single-line spectroscopic binary curves but have much lower amplitudes. From www.physics.sfsu.edu/~gmarcy/planetsearch/doppler.html.

Keeping in mind that $a = a_1$ is the semi-major axis relative to the center of mass, then, from (5), we have

$$\frac{a_1}{a_1 + a_2} = \frac{a_1}{a_0} = \frac{m}{M}$$

so that

$$V_r(\text{max}) = \frac{2\pi}{P} a_1 \sin i = \frac{2\pi}{P} \frac{m}{M} a_0 \sin i.$$

But from Kepler's Second Law

$$P^2 = \frac{4\pi^2}{GM} a_0^3 \rightarrow P = \frac{2\pi}{\sqrt{GM}} a_0^{3/2}.$$

Substituting into the expression for $V_r(\text{max})$

$$\Rightarrow V_r(\text{max}) = \sqrt{G} \frac{m}{\sqrt{M a_0}} \sin i.$$

In astronomer-friendly units

$$\frac{V_r}{\text{km s}^{-1}} \approx 30 \frac{m}{\sqrt{M a_0}} \sin i$$

where m, M are in solar masses and a_0 is in AUs.

Examples of velocity curves with planetary signatures are shown in Fig. 1.13. Over 100 planets have been detected with this procedure. Although this is the most sensitive technique for finding planets it has one major drawback. Only $m \sin i$ can be inferred, not m. It is therefore difficult to prove that any given candidate is a bona

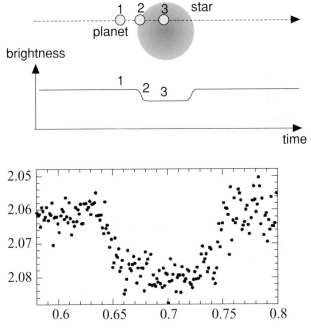

Fig. 1.14 Schematic of a transit, from Hans Deeg (2003), and actual data showing a transit of HD 209458, from Henry (2003). A transit occurs when the planet passes in front of a star, as shown. The resulting changes in the brightness of the star are shown in the case of HD 209458. The units are magnitude and time.

fide planet. Statistically, though, it would seem that at least some of the candidates are planets. The candidates are detailed further in the catalog of extrasolar planets.

1.4.3   The transit method

For stars oriented in just the right way, companions will transit (eclipse) in front of the star as a result of their orbital motion (Fig. 1.14). A light curve results very much like that of eclipsing binaries except that the depth of the eclipse is much smaller. This method is still in its infancy and suffers from the major disadvantage that it requires systems whose orbital planes are aligned with the observer's line of sight. Needless to say, statistics do not favor such an orientation. The first such searches have relied on observing known eclipsing binaries. It is not clear that binaries can support stable planetary orbits although that in itself is an interesting question that can be addressed by this method. The major advantage of the transit method is that it can, in principle, detect planets as small as the Earth, using current detector technology. The first possible detection of an extrasolar planet was reported by Charbonneau *et al.* (2000). Figure 1.14 shows a transit around the star HD 209458.

1.5 References

Charbonneau, D., Brown, T. M., Latham, D. W., Mayor, M. (2000) Detection of Planetary Transits Across a Sun-like Star, *Astrophysical Journal*, **529**, pp. L45–L48.

Deeg, H.-J. (2003) www.iac.es/proyect/tep/tephome.html.

Gatewood, G. (1996) *Bulletin of the American Astronomical Society*, **28**, 885.

Henry, G. (2003) http://schwab.tsuniv.edu/t8/hd209458/transit.html.

Torres, G., Stefanik, R. P., Latham, D. W. (1997) The Hyades Binary Finsen 342 (70 Tauri): A Double-lined Spectroscopic Orbit, the Distance to the Cluster, and the Mass–Luminosity Relation, *Astrophysical Journal*, **479**, pp. 268–78.

Torres, G., Stefanik, R. P., Latham, D. W. (1997) The Hyades Binaries theta 1 Tauri and theta 2 Tauri: The Distance to the Cluster and the Mass–Luminosity Relation, *Astrophysical Journal*, **485**, pp. 167–181.

1.6 Further reading

Goldstein, H. (1950) *Classical Mechanics*, Addison-Wesley, Reading, MA, USA.

Marcy, G. and Butler, P. (2003) http://exoplanets.org/science.html.

Schneider, J. (2003) www.obspm.fr/encycl/catalog.html.

Smart, W. M. (1977) *Spherical Astronomy*, Cambridge University Press, Cambridge, UK.

Chapter 2

Galaxy dynamics

Now that I have covered the two-body interaction in detail I turn my attention to more complex systems such as galaxies. Galaxies can be thought of as large systems of interacting objects such as stars and gas clouds. Since galaxies like the Milky Way contain of the order of 10^{11} stars it is not practical to analyze all possible two-body interactions in such a system. Instead, I treat the mass distribution of a galaxy as a continuous quantity and therefore determine the gravitational effects in a macroscopic fashion, that is, by considering the integral effects of matter on a test mass. I therefore begin with a discussion of forces and potentials arising from continuous but finite distributions of matter. In the discussions that follow I will relate forces and potential energy to test particles of *unit mass*. Much of the mathematical development in this chapter is patterned after that of Binney and Tremaine (1988).

2.1 Potentials of arbitrary matter distributions

Consider an incremental force $\delta \vec{F}(\vec{x})$ acting on a unit mass and arising from an infinitesimal mass element $\delta m(\vec{x})$. From Newton's law of universal gravitation (equation (1)) we have

$$\delta \vec{F}(\vec{x}) = G \frac{\vec{x}' - \vec{x}}{|\vec{x}' - \vec{x}|^3} \, \delta m(\vec{x}') = G \frac{\vec{x}' - \vec{x}}{|\vec{x}' - \vec{x}|^3} \rho(\vec{x}') \, \delta^3 \vec{x}'. \tag{43}$$

The total force arising from all mass elements of the mass distribution is obtained by integrating (43) so that

$$\vec{F}(\vec{x}) = \int \delta \vec{F}(\vec{x}) = G \int \frac{\vec{x}' - \vec{x}}{|\vec{x}' - \vec{x}|^3} \rho(\vec{x}') \, \delta^3 \vec{x}'. \tag{44}$$

Similarly, the gravitational potential of a continuous mass distribution can be expressed as

$$\Phi(\vec{x}) = -G \int \frac{\rho(\vec{x}')}{|\vec{x}' - \vec{x}|} \delta^3 \vec{x}'. \tag{45}$$

We can relate Φ to $\vec{F}$ by noting (see Jackson, 1998) that

$$\vec{\nabla}\left(\frac{1}{|\vec{x} - \vec{x}'|}\right) = \frac{\vec{x}' - \vec{x}}{|\vec{x}' - \vec{x}|^3}$$

and combining with (44) to yield

$$\vec{F}(\vec{x}) = \vec{\nabla} \int \frac{G\rho(\vec{x}')}{|\vec{x}' - \vec{x}|} \delta^3 \vec{x}' \tag{46}$$

or

$$\vec{F}(\vec{x}) = -\vec{\nabla}\Phi \tag{47}$$

which is the well-known result that the vector gravitational force is proportional to the gradient of the scalar gravitational potential.

Using these equations it can be shown (see Binney and Tremaine, 1988) that

$$\nabla^2 \Phi = 4\pi G\rho \qquad \text{Poisson's equation}$$

which reduces to

$$\nabla^2 \Phi = 0 \qquad \text{Laplace equation}$$

for the special case $\rho = 0$.

From Poisson's equation we can derive Gauss's theorem

$$4\pi G \int \rho \, d^3\vec{x} = \int \vec{\nabla}\Phi \cdot d^2\vec{S}. \tag{48}$$

Let us now put these equations to use.

2.2 Dynamics of thin disks

Galaxies such as the Milky Way are characterized by cylindrical disks that are very thin compared to the disk radii. To first order we can ignore their thickness completely for the sake of mathematical expediency. Consider therefore a circular disk of zero thickness characterized by some surface density distribution, $\Sigma(R)$, where R is the radial distance from the center of the disk (Fig. 2.1).

Since the disk is infinitely thin we can start with the Laplace equation rather than the Poisson equation and we should expect that the resulting solution is valid for all regions outside the disk (which means almost all regions). Since the disk has

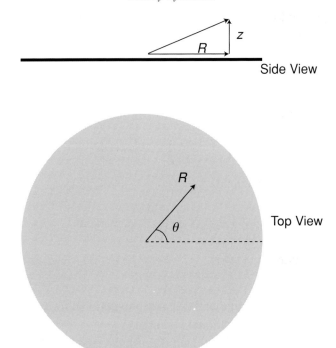

Fig. 2.1 Side and top views of an idealized galaxy disk. The relevant coordinates are R, θ and z.

cylindrical symmetry and since we want a solution for all 3-D space we begin by casting the Laplace equation in cylindrical coordinates.

$$\nabla^2 \Phi = \frac{1}{R}\frac{\partial}{\partial R}\left(\frac{1}{R}\frac{\partial \Phi}{\partial R}\right) + \frac{\partial^2 \Phi}{\partial z^2} = 0.$$

Separating variables, we look for a solution of the form

$$\Phi(R, z) = J(R)Z(z).$$

Substituting into the Laplace equation yields

$$\frac{1}{J(R)}\frac{1}{R}\frac{d}{dR}\left(R\frac{dJ}{dR}\right) = -\frac{1}{Z(z)}\frac{d^2Z}{dz^2} = -k^2,$$

which follows from the fact that the two independent quantities always add up to zero and this can only happen if the two terms are constrained to a constant value, k. Thus

$$\frac{d^2Z}{dz^2} - k^2 Z = 0$$

$$\frac{1}{R}\frac{d}{dR}\left(R\frac{dJ}{dR}\right) + k^2 J(R) = 0.$$

The first equation looks a lot easier to solve. In fact, we immediately see that its solution is given by

$$Z(z) = S\,e^{\pm kz} \tag{49}$$

where S is a constant. The second equation is a little trickier. We begin with the substitution, $u = kR$, which leads to

$$\frac{1}{u}\frac{d}{du}\left(u\frac{dJ}{du}\right) + J(u) = 0. \tag{50}$$

If we add the constraint that the solution should lead to a finite value of J at $u = 0$, we have

$$J(u) = J_0(u) \rightarrow J(kR) = J_0(kR). \tag{51}$$

The relevant solution is the cylindrical Bessel function of order zero. Combining the two solutions yields

$$\Phi(R, z) = S\,e^{\pm kz}J_0(kR). \tag{52}$$

Considering now a solution for a specific value of k, we have

$$\Phi_k(R, z) = e^{-k|z|}J_0(kR). \tag{53}$$

This solution has the desirable properties that $\Phi_k \rightarrow 0$ as $z \rightarrow \infty$ and $R \rightarrow \infty$. It also has the undesirable property that there is a discontinuity in the gradient of $\Phi(z)$ at $z = 0$. It does not satisfy Laplace's equation because matter is present there. We therefore need to correct the solution for the presence of the thin matter disk. Since the matter distribution is two dimensional, that is it forms a surface rather than a volume, it makes sense to use Gauss's theorem here. Since we are interested in finding the surface density that gives rise to the discontinuity in $\vec{\nabla}\Phi$

$$\frac{\partial\Phi_k}{\partial z} = -kJ_0(kR), \qquad \lim z \rightarrow 0^+$$

$$\frac{\partial\Phi_k}{\partial z} = kJ_0(kR), \qquad \lim z \rightarrow 0^-.$$

The integral of $\vec{\nabla}\Phi_k$ over a closed unit surface must equal $4\pi G\,\Sigma_k$ (48) so that

$$\Sigma_k(R) = -\frac{k}{2\pi G}J_0(kR) \tag{54}$$

is the surface density that we need.

Now we can determine the potential associated with the surface density. Using (53)

$$\Phi_k(R, z) = -e^{-k|z|}\frac{2\pi G\,\Sigma_k(R)}{k}. \tag{55}$$

Keep in mind that this is a special solution corresponding to a particular surface density Σ_k. To get the general solution for an arbitrary surface density Σ we need to allow for all possible values of k. If we can find a function $S(k)$ such that

$$\Sigma(R) = \int_0^\infty S(k)\Sigma_k(R)\,dk = -\frac{1}{2\pi G}\int_0^\infty S(k)J_0(kR)\,k\,dk \qquad (56)$$

then we will have

$$\Phi(R,z) = \int_0^\infty S(k)\Phi_k(R,z)\,dk = \int_o^\infty S(k)J_0(kR)\,e^{-k|z|}\,dk \qquad (57)$$

and the equations will be in terms of an arbitrary mass density $\Sigma(R)$.

The function $S(k)$ is the *Hankel transform* of $-2\pi G\Sigma$ and transforms in a similar fashion to Fourier transforms so that

$$S(k) = -2\pi G \int_0^\infty J_0(kR)\Sigma(R)R\,dR. \qquad (58)$$

Eliminating $S(k)$ from (57) and (58) yields

$$\Phi(R) = -2\pi G \int_0^\infty dk\, e^{-k|z|}\, J_0(kR) \int_0^\infty \Sigma(R')J_0(kR')R'\,dR'.$$

This is the solution we have been seeking. By stipulating the surface density we obtain the potential.

With the potential expressed in terms of the mass density it is now possible to determine the circular speed an object would have at any point on the disk ($z = 0$). Setting $z = 0$ we have

$$\frac{v^2}{R} = F = |\vec{\nabla}\Phi|$$

$$\Rightarrow v^2(R) = R\,|\vec{\nabla}\Phi| = R\left(\frac{\partial\Phi}{\partial R}\right)_{z=0} = -R\int_0^\infty S(k)J_1(kR)\,k\,dk. \qquad (59)$$

Equation (59) allows us to determine the rotation curve of a galaxy. Let us now do that for a galaxy like the Milky Way.

2.3   Rotation curves of disk galaxies

So far we have kept the surface mass density unspecified. Let us now consider a disk with a specific density distribution. Disk galaxies such as the Milky Way are characterized by optical surface brightnesses that decline exponentially with radius. Since it is generally believed that stars emit light in direct proportion to their mass, it is inferred that the radial mass distribution is also exponential. The surface mass density we will therefore assume is given by

$$\Sigma(R) = \Sigma_0 \, e^{-R/R_d}. \tag{60}$$

Substituting (60) into (59) and solving for the function $S(k)$, we have

$$S(k) = -\frac{2\pi G \Sigma_0 R_d^2}{[1 + (kR_d)^2]^{3/2}}. \tag{61}$$

To get the potential we simply insert (61) into (60) yielding

$$\Phi(R, z) = -2\pi G \Sigma_0 R_d^2 \int_0^\infty \frac{J_0(kR) \, e^{-k|z|}}{[1 + (kR_d)^2]^{3/2}} \, dk.$$

Since we are primarily interested in the dynamics of the thin disk itself we set $z = 0$ and solve the above integral – for solutions see Gradshteyn and Ryzhik (2000) and Abramowitz and Stegun (1964)

$$\Phi(R, 0) = -\pi G \Sigma_0 R (I_0(y) K_1(y) - I_1(y) K_0(y)) \tag{62}$$

where

$$y \equiv \frac{R}{2R_d}.$$

In (62), I_0 and I_1 are the modified Bessel functions of the first kind and of order 0 and 1 respectively. Similarly, K_0 and K_1 are modified Bessel functions of the second kind and of order 0 and 1 respectively.

To get the velocity of the disk as a function of radial distance R we need to differentiate (62) according to (59). Doing so, we obtain

$$v^2(R) = R \frac{d\Phi}{dR} = 4\pi G \Sigma_0 R_d y^2 \left[I_0(y) K_0(y) - I_1(y) K_1(y) \right]. \tag{63}$$

Equation (63) represents the rotation curve of a thin disk. It is a good representation of how we expect disk galaxies to rotate. A plot corresponding to (63), is shown in Fig. 2.2. Note the initial rise, the peak velocity and the subsequent $1/\sqrt{R}$ decline. The latter is known as *Keplerian rotation* because that is how the planets in the solar system behave.

2.3.1 Rotation curves of real spiral galaxies

Real spiral galaxies are not, of course, disks of zero thickness. Their disks have a finite width and they have spiral arms superimposed. In addition, the central region of many spiral galaxies is characterized by a spherical distribution of mass called the *bulge*. Conceptually, a real spiral galaxy looks something like Fig. 2.3. It is

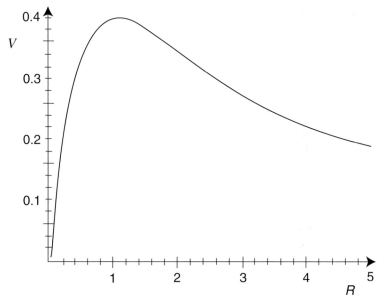

Fig. 2.2 Model rotation curves based on equation (63). The variables R and V are shown in relative units. Note the rigid body rotation evident at small R and the Keplerian fall-off at large R.

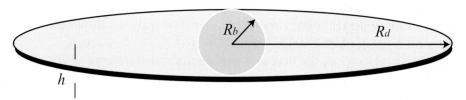

Fig. 2.3 The major parameters of a galaxy disk. The radius R_B of the central bulge and the radius R_d and thickness h of the disk are shown.

assumed that mass elements in the disk of a spiral galaxy have circular orbits about the center of the galaxy. The galaxy is taken to have a central bulge and a flat, circular disk.

The bulge

If the bulge has a constant density ρ_B, independent of radius, then

$$M(r) = \frac{4}{3}\pi r^3 \rho_B \qquad (r < R_B)$$

represents the mass distribution as a function of the radius, r, out to the edge of the bulge, $r = R_B$. Balancing centripetal and gravitational forces, we have

$$v^2(r) = \frac{GM(r)}{r} = \frac{4}{3}\pi\rho_B Gr^2 \qquad (r < R_B)$$

Thus, we see that

$$v(r) \propto r \qquad \text{and} \quad \dot\theta(r) = \text{constant.}$$

The linear velocity increases with r and the angular velocity is constant indicating rigid body rotation. This kind of rotation is observed in all spiral galaxy bulges and is consistent with the rotation of the inner disk as shown in Fig. 2.2.

The disk

Suppose that a point is reached where most of the disk is encompassed within an orbit of an outer mass element. Then we expect that mass element to respond as if it were orbiting a point mass. In the limit of large r, (63) yields

$$v(r) \propto \frac{1}{\sqrt{r}} \qquad \text{and} \quad \dot\theta \propto r^{-3/2}$$

consistent with $M(r) = $ constant and Keplerian rotation. These dependencies are indicative of *differential rotation*, a characteristic shared by many astronomical bodies that are not solid or rigid.

Thus, it is expected that in the outer-most regions of a spiral galaxy there should be a turnover where $v(r)$ begins to decline according to Keplerian orbital motion. *Such a turnover is rarely observed in spiral galaxies!* In fact, most rotation curves appear to be flat, i.e. $v(r) = $ constant. Why is that? Well, if we assume that Kepler's laws are valid on kiloparsec scales then we must conclude that the true mass distribution is different from that which is actually visible. The unseen mass may not radiate but it still provides a gravitational potential that any mass element must respond to.

The halo

Let us therefore suppose that there is a massive, spherical and invisible halo of material surrounding a spiral galaxy such that

$$M(r) = \frac{4}{3}\pi r^3 \rho_h \qquad (r < R_h)$$

and let us suppose that the halo dominates all other components in terms of total mass. It turns out that such massive halos are gravitationally stable if $\rho_h \propto 1/r^2$ in which case

$$M(r) \propto r$$
$$v^2(r) \propto \frac{M(r)}{r} = \text{constant}$$
$$\Rightarrow v(r) = \text{constant.}$$

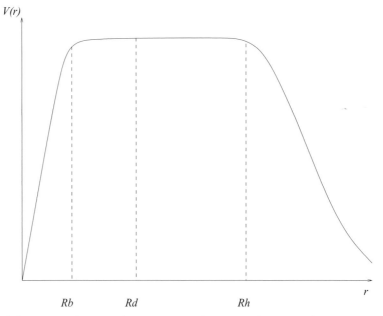

Fig. 2.4 Hypothetical rotation curve showing a massive halo. If a massive dark matter halo did exist and we had the means to measure rotation into the halo, the resulting rotation curve might look like the one shown.

We see that massive halos of dark matter can *flatten* rotation curves of spiral galaxies. This fact is the main argument that individual galaxies are surrounded by dark matter. Dark matter and the rotation curves of galaxies are discussed in greater detail in Chapter 14. In Sections 2.4.3 and 3.2.3, I discuss evidence for dark matter in clusters of galaxies. If dark matter halos exist then the true rotation curves of galaxies might look similar to the sketch in Fig. 2.4.

2.4 *N*-body gravitational systems

Consider a system of many particles (galaxies) that interact only via gravity.

2.4.1 *Equation of motion*

We can set up an equation of motion using the law of universal gravitation (equation (1))

$$m_i \ddot{\vec{r}}_i = -G \sum_{j \neq i} \frac{m_i m_j}{|\vec{r}_i - \vec{r}_j|^3} (\vec{r}_i - \vec{r}_j). \tag{64}$$

Taking the product of $\dot{\vec{r}}_i$ with (64) and summing over all i, the left-hand side becomes

$$\sum_i m_i \ddot{\vec{r}}_i \cdot \dot{\vec{r}}_i = \frac{d}{dt} \sum_i \frac{m_i \dot{r}_i^2}{2} = \frac{d}{dt} E_k \tag{65}$$

while the right-hand side becomes

$$-G \sum_{i,j} \frac{m_i m_j}{|\vec{r}_i - \vec{r}_j|^3} (\vec{r}_i - \vec{r}_j) \cdot \dot{\vec{r}}_i$$

$$= -G \sum_{i,j} \frac{m_i m_j}{|\vec{r}_j - \vec{r}_i|^3} (\vec{r}_j - \vec{r}_i) \cdot \dot{\vec{r}}_j$$

$$= \frac{1}{2} G \sum_{i,j} m_i m_j \frac{d}{dt} \frac{1}{|\vec{r}_i - \vec{r}_j|} = -\frac{d}{dt} E_G.$$

Equating the two sides we see that

$$\frac{d}{dt}(E_k + E_G) = \frac{d}{dt} E_{tot} = 0.$$

This equation is not only a statement of energy conservation but is also useful as an equation of motion for large numbers of bodies under the influence of gravity.

2.4.2 The Virial theorem

The Virial theorem is a valuable tool for studying static, non-evolving (relaxed) systems such as stars (systems of gas particles), gas clouds, star clusters, galaxies and galaxy clusters. The Virial theorem can be used to study large structures in the Universe as now described.

Let us go back to (64) and dot-multiply the left-hand side of the expression with $\vec{r}_i$ and then sum over all i

$$\sum_i m_i \ddot{\vec{r}}_i \cdot \vec{r}_i = \sum_i m_i \left[\frac{d^2}{dt^2} \left(\frac{r_i^2}{2} \right) - \dot{r}_i^2 \right]$$

$$= \frac{d^2}{dt^2} \sum_i \frac{m_i r_i^2}{2} - \sum_i m_i \dot{r}_i^2 = \frac{d^2}{dt^2} I - 2E_k,$$

where I is the moment of inertia of the system. For a stationary system (one that is relaxed) the moment of inertia does not change (on average) with time. Thus

$$\sum_i m_i \ddot{r}_i r_i = -2E_k. \tag{66}$$

Repeating the same operation on the right-hand side of (64) we get

$$-\sum_{ij} m_i m_j G \frac{(\vec{r}_i - \vec{r}_j) \cdot \vec{r}_i}{|\vec{r}_i - \vec{r}_j|^3}$$

$$= -\sum_{ij} m_i m_j G \frac{(\vec{r}_j - \vec{r}_i) \cdot \vec{r}_j}{|\vec{r}_i - \vec{r}_j|^3}$$

$$= -\frac{1}{2} \sum_{ij} \frac{m_i m_j G}{|\vec{r}_i - \vec{r}_j|} = E_G. \tag{67}$$

Equating (66) and (67) yields the relation that defines the theorem

$$E_k = -\frac{1}{2} E_G = \frac{1}{2} |E_G|. \tag{68}$$

$$\Rightarrow \quad \text{the Virial theorem}$$

According to the Virial theorem the total energy of a stationary system is

$$E_T = E_k + E_G = \frac{1}{2} E_G.$$

A stationary system is one in which no significant dynamical evolution is taking place.

2.4.3 Clusters of galaxies

Let us consider now a cluster of galaxies and the application of the Virial theorem to such a cluster. The kinetic energy can be obtained by summing over the kinetic energies of the individual galaxies in the cluster.

$$E_k = \sum \frac{M_i V_i^2}{2} = M_{tot} \frac{\langle V^2 \rangle}{2}. \tag{69}$$

The space velocities of the galaxies cannot be observed directly but their radial velocities, V_{rad} can. For a sufficiently large number of galaxies, there is a relationship between the average radial velocities of the galaxies and their average space velocities. Assuming isotropic orbits, it is given by

$$\langle V_{rad}^2 \rangle = \langle V^2 \rangle / 3.$$

The self-potential or gravitational energy can be expressed as

$$E_G = -\frac{G M_{tot}^2}{R_c}, \tag{70}$$

where R_c is the radius of the cluster. By combining (68), (69) and (70) we obtain

Fig. 2.5 The Hercules cluster of galaxies. Note the mix of elliptical and spiral galaxies. From: http://antwrp.gsfc.nasa.gov/apod/ap980827.html. **Credit:** Dr. Victor Andersen (University of Alabama, KPNO).

an expression for estimating the total mass of the cluster.

$$M_{tot} = \frac{3\langle V_{rad}^2\rangle R_c}{G}.$$

The radius of the cluster, R_c, can be estimated if the distance to the cluster is known (obtained from the Hubble relation). Masses of clusters, obtained in the above manner, have led to estimates of the *mass-to-light ratio*, that is

$$\frac{M_{tot}}{L} \approx 200\frac{M_\odot}{L_\odot}.$$

This is a much higher number than is typically found for individual galaxies indicating that there must be a lot of matter whose gravitational influence is felt but is otherwise invisible. This ratio of mass to light can be used to estimate the average density of the Universe (see Chapter 3).

There are uncertainties in these analyses that must be kept in mind.

(i) The characteristic radius of the cluster is determined from the distribution of visible matter and may not be representative of the true matter distribution.
(ii) The outer parts of the cluster may not be static in which case, $d^2 I/dt^2$ may not be 0.
(iii) The mass-to-light ratio may be a function of radius within the cluster.
(iv) Newtonian gravitation may be invalid on large scales.

Examples of well-known galaxy clusters are shown in Figs. 2.5 to 2.7.

Fig. 2.6 The Virgo cluster of galaxies. The field is dominated by a pair of giant elliptical galaxies. From: http://www.seds.org/messier/more/virgo.html.

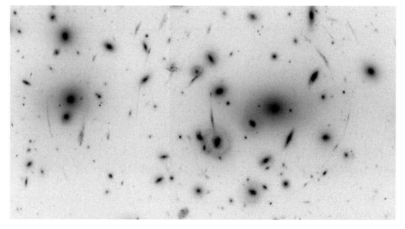

Fig. 2.7 Abell 2218: A Galaxy Cluster Lens. The foreground galaxy cluster is refracting, gravitationally, the light from distant background objects. From: http://antwrp.gsfc.nasa.gov/apod/ap950710.html **Picture Credit:** NASA, HST, WFPC2, W. Couch (UNSW).

2.5 References

Abramowitz, M. and Stegun, I. A. (1964) *Handbook of Mathematical Functions with Formulas, Graphs, and Mathematical Tables, National Bureau of Standards Applied Mathematics Series, vol. 55*, US Government Printing Office, Washington, DC, USA.

Binney, J. and Tremaine, S. (1988) *Galactic Dynamics*, Princeton University Press, Princeton, NJ, USA.

Gradshteyn, I. S. and Ryzhik, I. M. (2000) *Table of Integrals, Series, and Products*, 6th edn (translated from Russian by Scripta Technika), Academic Press, New York, NY, USA.

2.6 Further reading

Bertin, G. (2000) *Dynamics of Galaxies*, Cambridge University Press, New York, NY, USA.

Binney, J. and Merrifield, M. (1998) *Galactic Astronomy*, Princeton University Press, Princeton, NJ, USA.

Combes, F., Mamon, G. A., Charmandaris, V. (Eds) (2000) *Dynamics of Galaxies: From the Early Universe to the Present*, Astronomical Society of the Pacific, San Francisco, CA, USA.

Jackson, J. D. (1998) *Classical Electrodynamics*, John Wiley and Sons, New York, NY, USA.

Chapter 3
Cosmic expansion and large scale structure

3.1 The expansion of the Universe

The expansion of the Universe can be modeled in a very simple way if we assume *uniform expansion* and that matter dominates the dynamics of the expansion (in other words, purely gravitational dynamics). Let us therefore consider a sphere expanding uniformly such that

$$\vec{V} = H\vec{r}$$

where $\vec{V}$ is the velocity of expansion and $\vec{r}$ is the radius at which the velocity is measured (Fig. 3.1). We also assume complete spherical symmetry. The kinetic energy of the expansion can be written as

$$E_k = \int \frac{\rho v^2}{2} \, d^3r = \frac{4\pi \rho H^2}{2} \int_0^R r^4 \, dr = \frac{2\pi}{5} \rho H^2 R^5.$$

The gravitational potential energy can be expressed as

$$E_G = -G \int \frac{M(r)\rho}{r} \, d^3r = -G \int \left(\frac{4\pi}{3} r^3 \rho \right) \frac{\rho}{r} \, d^3r$$

$$\Rightarrow E_G = \frac{(4\pi)^2}{15} \rho^2 G R^5.$$

We see that

$$M_R \propto \rho R^3$$

$$R \propto \left(\frac{M_R}{\rho} \right)^{1/3}$$

$$E_G \propto \rho^{1/3} M^{5/3} \to 0 \quad \text{as} \quad \rho \to 0.$$

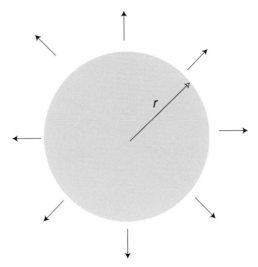

Fig. 3.1 Cosmic expansion according to Newtonian mechanics. The expansion is assumed to be uniform and isotropic.

The total energy in the expansion can now be obtained

$$E_T = E_k \left(1 + \frac{E_G}{E_k} \right) = E_k \left(1 - \frac{8\pi\rho G}{3H^2} \right) \equiv E_k \left(1 - \frac{\rho}{\rho_c} \right) \equiv E_k (1 - \Omega)$$

where

$$\rho_c = \frac{3H_0^2}{8\pi G} = 10^{-29} \text{g cm}^{-3} \left(\frac{H_0}{70 \, \text{km s}^{-1} \, \text{Mpc}^{-1}} \right)^2$$

$$= 6 \times 10^{-6} m_p \, \text{cm}^{-3} \left(\frac{H_0}{70 \, \text{km s}^{-1} \, \text{Mpc}^{-1}} \right)^2 ,$$

where H_0 is known as the Hubble constant and represents the present rate of expansion.

We know from Earth-based and space-based observations that

$$H_0 \approx 70 \, \text{km s}^{-1} \, \text{Mpc}^{-1} = 2.3 \times 10^{-18} \, \text{s}^{-1} = (14 \, \text{billion years})^{-1}$$

to within about 30%. Note that H_0 has units of inverse time so that $1/H_0$ represents a timescale for the expansion,

$$\Rightarrow t_0 \approx H_0^{-1} = 1.4 \times 10^{10} \, \text{years}.$$

Using these relationships it is possible to determine the radius of the Universe as a function of time. The exact shape of the function $R(t)$, depends on the quantity Ω, the density of the Universe. Figure 3.2 illustrates the cases, $\Omega < 1$, $\Omega = 1$ and $\Omega > 1$. If $\rho < \rho_c$ ($\Omega < 1$) then $E_T > 0$ so that as $\rho \to 0$, $E_G \to 0$ and E_k remains > 0. Thus expansion continues without termination.

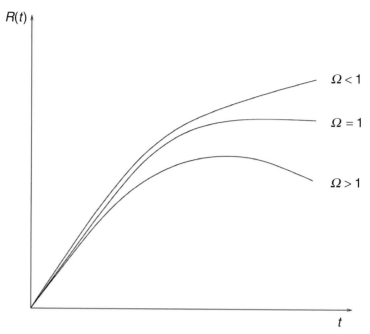

Fig. 3.2 Possible expansion scenarios. In the Newtonian model of the Universe, the future evolution of the universe is determined by the parameter Ω. An open, flat or closed universe is possible, depending on the mass density of the Universe.

If, on the other hand, $\rho > \rho_c$ $(\Omega > 1) \Rightarrow E_T < 0$. Then a time is eventually reached when $E_k = 0$ and $E_T = E_G$. Without any kinetic energy left in the expansion the gravitational term takes over and the Universe re-collapses to a point.

Empirical determinations of Ω have narrowed its value to

$$\Omega \approx 0.3.$$

This value of Ω suggests that the Universe is open. However, recent observations suggest that Ω is not the only parameter that characterizes the expansion of the Universe.

3.1.1 *The cosmological constant*

Until recently, the above simple picture provided a good qualitative picture of how the Universe expands. However, recent observations of galaxy redshift–distance relations have revealed that the Universe may be actually accelerating as opposed to decelerating. How can this be?

In order for the Universe to accelerate, there must be an effective "force" that is countering gravity. Current speculation centers on the "cosmological constant". The cosmological constant, Λ, is a term in Einstein's general relativity equations that

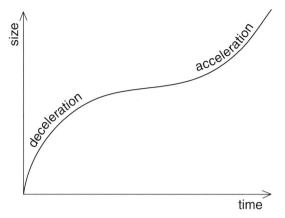

Fig. 3.3 The distance $R(t)$ for an expanding Universe with a nonzero Λ. The inflection point corresponds to the present epoch. The past expansion has been dominated by matter, future expansion will be controlled by Λ.

describes the expansion of the Universe. Einstein had placed the term there when the Universe was believed to be static. A re-examination of those equations leads to the following equation of motion for the Universe

$$\frac{3\ddot{R}}{R} = -\frac{c}{2R^3} + \Lambda.$$

The behavior of this function is interesting. For small values of R (early Universe) the solution is similar to that which we derived for the Newtonian case. For large R, the first term on the right-hand side becomes insignificant, leading to

$$\frac{3\ddot{R}}{R} = \Lambda$$

which has the solution, for large R

$$R \propto e^{\sqrt{3\Lambda}t/3}.$$

The expansion of the Universe becomes exponential.

A plot of the formal (full) solution is sketched in Fig. 3.3. Note that we are near the inflection point. The expansion of the Universe has been dominated by gravity in the past and will be dominated by the cosmological constant in the future. The physical meaning of Λ is not fully known but current theories suggest that vacuum energy is the source of the "repulsive force".

The geometry of the Universe is now determined by $\Omega = \Omega_\Lambda + \Omega_M$ where Ω_M replaces the Ω determined earlier for a purely Newtonian Universe. Current estimates of the values of Ω_Λ and Ω_M are based on plots such as the one shown in Fig. 3.4.

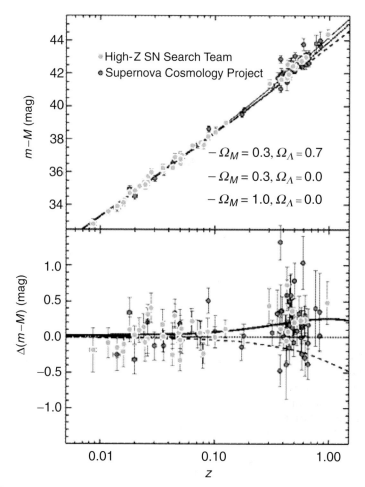

Fig. 3.4 The observations that led to a nonzero Λ. The top plot shows the distance moduli (relating to actual distance) plotted as a function of redshift, z. The observed points are shown as dots, along with best fit model curves utilizing different combinations of Ω_Λ and Ω_M. The differences in the fits are more clearly seen in the bottom figure, which results from subtracting the $\Omega_\Lambda = 0$, $\Omega_M = 0.3$ model curve from the data points. The excess of positive data points near $z = 0.5$ is best fit by the $\Omega_\Lambda = 0.7$, $\Omega_M = 0.3$ model. From: Perlmutter, S. *et al.*, The Supernova Cosmology Project, 1999, ApJ, 517, pp. 565–586.

Inspection of Fig. 3.4 shows that the best fit solution yields, $\Omega_M \approx 0.3$ and $\Omega_\Lambda \approx 0.7$. The sum of the two terms is consistent with $\Omega = 1$. Thus, it appears that we live in a flat Universe.

There is still considerable uncertainty in these estimates, as shown in Fig. 3.5. Future projects, such as the Supernova Acceleration Probe (SNAP), are designed to bring down the uncertainties, as illustrated in Fig. 3.5.

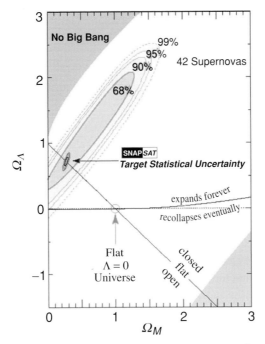

Fig. 3.5 The values of Ω_Λ and Ω_M. These results from the Supernova Cosmology Project show the best fit contours in the Ω_Λ–Ω_M plane. The uncertainty is displayed in terms of confidence levels. Although the uncertainties are large they are consistent with a nonzero Ω_Λ. Future projects such as SNAP should greatly reduce the uncertainties shown hypothetically in the plot. From: Perlmutter, S. *et al.*, The Supernova Cosmology Project, 1999, ApJ, 517, pp. 565–586.

3.2 Large-scale cosmic structure

I now discuss the characteristics of large scale structure and gravitational dynamics on large scales. Expanding upon the preceding discussion I relate the current large scale structure to the conditions in the early Universe that led to the formation of galaxies. Figure 3.8 shows the distribution of galaxies for a specific "slice" of the Universe, indicating that galaxies are not distributed randomly in space.

3.2.1 Overview

The following observations describe the main characteristics of large scale structure in the modern Universe.

- Stars are distributed uniformly on scales of ≈ 10 kpc and mass concentrations of $\approx 10^{11} M_\odot$ which we call galaxies. We can think of galaxies as test particles that trace out the structure of the Universe. Even "empty" regions of the sky contain countless galaxies if we look hard enough (see Fig. 3.9).

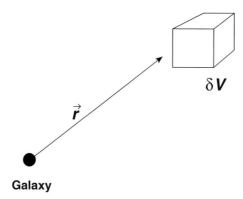

Galaxy

Fig. 3.6 A volume element near a galaxy. The definitions of $\vec{r}$ and δV used in galaxy correlation studies are as shown.

- The distribution of galaxies is non-uniform.
- Galaxies trace out structures we call clusters, superclusters and voids (see Fig. 3.8).
- The big question is: How did these structures form and how do they relate to the formation of galaxies?

Before addressing the last question, let us first describe a quantitative measure of cosmic structure, the correlation function of galaxies, a tool that can help illustrate the role of dark matter in large scale structure. The discussion is motivated by the developments of Narlikar (1993).

3.2.2 *Correlation functions of galaxies*

Consider an elemental volume of space δV, a distance r from a given galaxy (Fig. 3.6). What is the probability of finding a galaxy in that volume? The elemental probability is given by

$$\delta p = \bar{n} \left\{ 1 + \zeta(r) \right\} \delta V$$

where $\bar{n}$ is the local mean density of galaxies. The function $\zeta(r)$ represents the deviation from a uniform distribution. Positive means an enhancement of galaxies while a negative means a depletion.

Galaxy counts can be used to determine the shape of the function, $\zeta(r)$. Such measurements yield

$$\zeta(r) \propto \left(\frac{r}{r_0} \right)^{-\gamma}$$

where $r_0 = 5h_0^{-1}$ Mpc, $\gamma \approx 1.8$ and $h_0 = H_0/75$. A major result is that γ appears to be scale invariant.

The correlation functions describe the distributions of visible matter. But we have some hints that dark matter plays an important dynamical role. In order to build a theory that describes galaxy formation and the formation of large scale structure we need to take into account the possible properties of dark matter.

3.2.3 Dark matter and large-scale structure

Dark matter is obviously difficult to detect. One approach is to determine the mass-to-light ratios of the various structures in the Universe. We begin by defining the mass-to-light ratio as

$$\eta = \frac{M(M_\odot)}{L(L_\odot)}.$$

Galaxies

It is possible to estimate the "visible" mass of an individual galaxy from its rotation curve via

$$\frac{v^2(r)}{r} = \frac{GM(r)}{r^2}$$

$$\Rightarrow M(r) = \frac{rv^2(r)}{G}$$

where $v(r)$ is the rotational velocity at radius r. Since most galaxy rotation curves do not fall off, the resulting mass is a lower limit. Such measurements yield

$$\eta = (9 \pm 1)h_0 \qquad \text{Spirals}$$
$$\eta = (10 \pm 2)h_0 \qquad \text{Ellipticals.}$$

For comparison, the Sun has $\eta = 1$, an M5 star has $\eta = 20$ and an O star has $\eta = 2 \times 10^{-4}$.

Galaxy groups

Consider a pair of galaxies, moving non-relativistically, as shown in Fig. 3.7. Typically, $\vec{v} = \vec{v}_H + \vec{v}_p$, where $\vec{v}_H$ is the velocity resulting from the Hubble flow and $\vec{v}_p$ is the velocity arising from peculiar motions caused by the local gravitational potential. The velocity of object G can be similarly expressed so that

$$\Delta \vec{V} = \vec{v}_G - \vec{v}_g = \Delta \vec{v}_H + \Delta \vec{v}_p = H \Delta \vec{r}_H + H \Delta \vec{r}_p$$

where $\Delta \vec{r}_H + \Delta \vec{r}_p$ represents the apparent radial component of the separation vector. The total apparent separation is given by

$$\Delta \vec{r} = \Delta \vec{r}_H + \Delta \vec{r}_p + \vec{\sigma} = \vec{\pi} + \vec{\sigma}$$

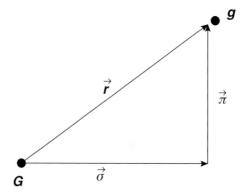

Fig. 3.7 Components of galaxy motion. The motion of galaxy g relative to galaxy G has two components.

which now includes the transverse component, $\vec{\sigma}$. The true separation is given by

$$\Rightarrow \Delta\vec{r}_t = \Delta\vec{r}_H + \vec{\sigma}.$$

Thus a plot of π versus σ for a large number of galaxies results in a distorted correlation where the points clump vertically, parallel to the radial axis.

Such plots can be used to yield $\langle \Delta r_p \rangle$ and $\Delta\vec{v}_p$. Typically

$$\langle v_p^2 \rangle^{1/2} \approx (600 \pm 250) \text{ km s}^{-1}.$$

This information can be used to estimate masses of galaxy groups as follows. Recall from the Virial theorem that

$$v_i^2 \approx \frac{GMN_i}{R}.$$

where v_i is the "peculiar" velocity one would expect of the ith galaxy having N_i neighbors of average mass, M, at a distance R. The N_i can be estimated from the correlation function, $\zeta(r)$ for a 5 Mpc volume around our Galaxy where

$$N = \bar{n} \int_0^{5\,\text{Mpc}} [1 + \zeta(r)] \, d^3r \approx 42$$

and $\bar{n} = 0.03 h_0^3$ Mpc^{-3} (for bright galaxies). Setting $R = r_0$, $N = \langle N_i \rangle$, the result is

$$M \approx 5 \times 10^{12} \, h_0^{-1} \, M_\odot$$

for a volume of space with $r_0 = 5$ Mpc. Combining this result with empirical estimates of the luminosity, L, for the same volume, yields

$$\eta \approx (500 \pm 200) h_0.$$

An image of Stephan's Quintet, a group of galaxies, is shown in Fig. 3.10.

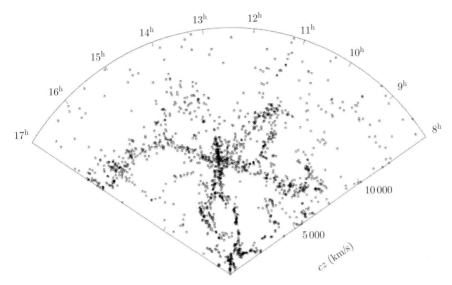

Fig. 3.8 Structure in redshift space. A slice of the Universe is shown. Each dot represents a galaxy whose RA coordinate and redshift have been measured. The resulting plot is a plane that shows the distribution of galaxies in depth. It is seen that galaxies are mainly found in large filamentary structures, which can be thought of as walls separating large empty regions known as voids. From: http://nedwww.ipac.caltech.edu/level5/March01/Salzer/Figures/figure2.jpg.

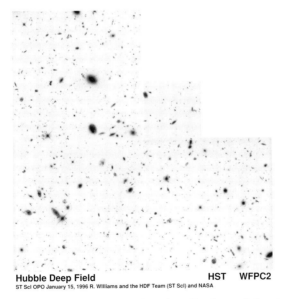

Fig. 3.9 Deep HST field of an "empty" region. Regions of the sky that appear empty to the naked eye are filled with galaxies, as shown in this famous Hubble deep field. From: http://arcturus.mit.edu/gallery/gifs/HDF-MosaicHalf.jpg.

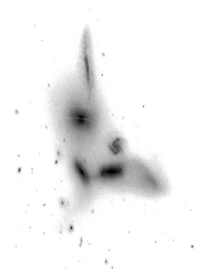

Fig. 3.10 Stephan's Quintet – an interacting group of galaxies. This small group of galaxies is too small to be classified as a cluster though the five galaxies are clearly interacting gravitationally. From: http://antwrp.gsfc.nasa.gov/apod/ap030124.html. Credit: J. English (U. Manitoba), C. Palma (PSU), *et al.*, NASA.

Clusters of galaxies

For well-defined clusters such as Coma

$$\eta \approx 300 \approx 30\eta_G.$$

An image of the Coma cluster is shown in Fig. 3.11.

The local supercluster

Similar measurements of the Virgo supercluster (we sit in its outskirts at $d \approx 11\, h_0^{-1}$ Mpc) yield

$$\eta \approx (80 \pm 30)h_0.$$

Thus, we see that light-to-mass ratios of all structures larger than galaxies are much greater than those of the individual galaxies.

Studies such as these have been used to estimate the amount of matter in the Universe. These estimates suggest that

$$\Omega_G \approx 0.3.$$

Fig. 3.11 The Coma cluster of galaxies. Coma is one of the richest galaxy clusters known. Most of the objects seen in this image are galaxies. From: http://www.noao.edu/image_gallery/html/im0118.html. Credit: "NOAO/AURA/ NSF".

3.2.4 Hot and cold dark matter

The particle properties of dark matter determine what structural scales are formed in the early Universe. Thus, it is important to review possible types of dark matter. If the larger values of η prove to be correct, then the gravitational dynamics of the Universe may have been driven entirely by dark matter. The visible matter may simply have followed but not influenced the overall dynamics of the Universe.

Measurements of the microwave background fluctuations suggest that the fluctuations are too small for visible matter to form the structures we see now. However, if particles were present which did not interact well with photons or baryons, their properties could well have determined these structural scales because they are capable of dominating the gravitational dynamics of the Universe.

Suppose that another form of matter existed, some sort of dark matter. Initially, the particles that make up the dark matter would be relativistic because of the rapid expansion of the Universe. Relativistic particles fill and smooth out the matter distribution in a horizon sphere of size $2ct_*$

$$R_* \approx ct_* \tag{71}$$

so that at epoch t_*, the largest mass structure is given by

$$\mathcal{M} = \frac{4}{3}\pi R_*^3 \left(\frac{\epsilon}{c^2}\right) \tag{72}$$

where ϵ is the energy density of the sphere.

As the Universe cools, the particles become sub-relativistic when the temperature drops to

$$T_* = \frac{m_* c^2}{k} \tag{73}$$

where m_* is the mass of the particle. The epoch corresponding to a temperature T_* is given by

$$t_* \propto T_*^{-2} \tag{74}$$

and the energy density by

$$\epsilon_* \propto (kT_*)^4. \tag{75}$$

Combining (72), (73), (74) and (75)

$$\Rightarrow \mathcal{M} \propto m_*^{-2}.$$

The exact solution is given by

$$\mathcal{M} \approx 10^{15} \left(\frac{30\,\text{eV}}{m_*}\right)^2 M_\odot \tag{76}$$

where m_* is in electronvolt units. Thus, the mass of the dark matter particle has a direct bearing on how small a fluctuation can exist in the matter-dominated Universe.

For a particle with $mc^2 \approx 30\,\text{eV}$, $\rightarrow \mathcal{M} \approx 10^{15}\,M_\odot$. This is one prediction of *Hot Dark Matter*.

On the other hand, for a particle with $mc^2 \approx 1\,\text{MeV}$, $\rightarrow \mathcal{M} \approx 10^6\,M_\odot$. This is the prediction of *Cold Dark Matter*.

3.2.5 *The Jeans' mass and gravitational stability*

Primordial density fluctuations are constrained to having a minimum mass because the conditions at decoupling are such that the thermal pressure of matter can balance gravitational collapse (Chapter 5). We now calculate the minimum mass based on what we know about primordial fluctuations.

In terms of the total energy we have the following three cases that define dynamical stability

$$2E_k + E_G = 0 \qquad \text{static equilibrium}$$

$$E_k < -\frac{1}{2}E_G \qquad \text{collapse}$$

$$E_k > -\frac{1}{2}E_G \qquad \text{expansion.}$$

In the case of galaxy clusters the kinetic energy refers to the motion of individual galaxies. In the case of a clump of gas, it refers to the motion of the individual gas particles, the atoms. Thus, for a parcel of gas, assumed to be ideal, we can rewrite the condition for collapse as

$$E_k = \frac{3}{2}NkT < -\frac{1}{2}E_G$$

$$\Rightarrow 3NkT < \frac{GM^2}{\langle R \rangle} \qquad \text{Jeans' Condition.} \tag{77}$$

From the Jeans' condition we see that there is a minimum mass below which the thermal pressure prevents gravitational collapse.

$$M_{min} = M_J = \left(\frac{3NkT\langle R \rangle}{G} \right)^{1/2}.$$

The number of atoms corresponding to the Jeans' mass is given by

$$N = \frac{M_J}{m_p \mu} \tag{78}$$

where μ is the mean molecular weight of the gas and m_p is the mass of the proton. In terms of the mass density, ρ,

$$M_J = \frac{4\pi}{3}\rho\langle R \rangle^3. \tag{79}$$

Combining (77), (78) and (79)

$$M_J = \rho^{-1/2}T^{3/2}\left(\frac{5k}{G\mu m_p} \right)^{3/2}\left(\frac{3}{4\pi} \right)^{1/2}.$$

As expected, high density favors collapse while high temperature favors larger Jeans' mass. In units favored by astronomers this condition becomes

$$M_J \approx 45\ M_\odot T_k^{3/2}n_{cm^{-3}}^{-1/2}. \tag{80}$$

At the era of decoupling, $T_k \approx 3000$ and $n \approx 6 \times 10^3$. Inserting into equation (80) yields

$$M_J(t_D) \approx 10^5 \, M_\odot.$$

Thus, the smallest possible mass capable of collapse at the time of decoupling was $10^5 \, M_\odot$ (assuming that those scales exist). That is about the mass of a globular cluster today. Nothing smaller could have formed. By contrast, in the interstellar medium of our Galaxy where $T \approx 50$ K and $n \approx 10^3$ the Jeans' mass is $\approx 500 \, M_\odot$. Now, far smaller parcels of gas can collapse. That is why molecular clouds can collapse and form individual stars. The Jeans' mass at recombination is an important constraint on models attempting to explain how galaxies formed.

3.2.6 Possible models of structure formation

Any theory that attempts to explain large-scale structure must account for:

(a) The presence of clusters, superclusters and voids.
(b) The $r^{-1.8}$ scale-invariant correlation law.
(c) The minimum mass at decoupling.
(d) The role of dark matter.

There are two major competing models that attempt to explain the formation of the large scale structure that we now see. One model begins with the largest structures forming first and the smallest structures forming last. The other proposes that smallest structures form first which then coalesce to form larger structures. The two models are summarized below.

(A) The top-down scenario

- Largest structures form first because the expanding *hot dark matter* smoothes out all smaller scales.
- Other, smaller structures result from subsequent fragmentation of the largest structures.
- Galaxies form last in this fragmentation process.
- The fragmentation follows the $r^{-1.8}$ law.
- Hot dark matter is needed to form the large structures from the primordial fluctuations.

(B) The bottom-up scenario

- Objects with $\mathcal{M} \approx 10^{5-6} \, M_\odot$ form first because that is the minimum mass allowed according to the Jeans' criterion.
- Larger structures result from gravitational interactions of the smaller structures.
- Galaxies form before larger-scale structure.
- The $r^{-1.8}$ structure law is achievable.
- Structure exists on small scales because of the dominance of cold dark matter (CDM).

In recent years, the realization that the cosmological constant may play a role in the expansion of the Universe, has led to the development of models that incorporate nonzero values of Λ. The favored models now are those that utilize CDM and nonzero Λ. A more detailed discussion of the early Universe (the time before decoupling of matter and radiation) is presented in Chapter 5.

3.3 References

Narlikar, J. V. (1993) *Introduction to Cosmology*, Cambridge University Press, Cambridge, UK.

Perlmutter, S. *et al.* (1999) Measurements of Omega and Lambda from 42 High-Redshift Supernovae, *Astrophysical Journal*, **517**, pp. 565–86.

3.4 Further reading

Coles, P. and Ellis, G. F. R. (1997) *Is the Universe Open or Closed?: The Density of Matter in the Universe*, Cambridge University Press, New York, NY, USA.

Combes, F., Mamon, G. A., Charmandaris, V. (Eds) (2000) *Dynamics of Galaxies: From the Early Universe to the Present*, Astronomical Society of the Pacific, San Francisco, CA, USA.

Dekel, A. and Ostriker, J. (Eds) (1999) *Formation of Structure in the Universe*, Cambridge University Press, Cambridge, UK.

Fairall, A. P. (Ed). (1998) *Large-Scale Structures in the Universe*, John Wiley and Sons, New York, NY, USA.

Harrison, E. (2000) *Cosmology: The Science of the Universe*, Cambridge University Press, Cambridge, UK.

Harwitt, M. (1991) *Astrophysical Concepts*, Springer-Verlag, New York, NY, USA.

Liddle, A. R. and Lyth, D. H. (2000) *Cosmological Inflation and Large Scale Structure*, Cambridge University Press, Cambridge, UK.

Peacock, J. A. (1999) *Cosmological Physics*, Cambridge University Press, Cambridge, UK.

Peebles, P. J. E. (1993) *Physical Cosmology*, Princeton University Press, Princeton, NJ, USA.

Rees, M. (2000) *New Perspectives in Astrophysical Cosmology*, Cambridge University Press, Cambridge, UK.

Part II

Statistical mechanics

There are about 10^{80} particles in the Universe. Most of them are concentrated in stars though some can be found in interstellar and intergalactic space. At the centers of stars, particle densities are sufficiently high to allow nuclear reactions to take place. The energy liberated by these reactions heats the gas that makes up the star to the point where the gas pressure balances the gravitational pressure leading to hydrostatic equilibrium and long-term stability. Stars like the Sun are stable for about 10^{10} years, a large fraction of the age of the Universe.

The radiation produced by the nuclear reactions achieves a near equilibrium with the gas particles in the stellar interior, leading to local (but not global) thermodynamic equilibrium. The presence of local thermodynamic equilibrium (LTE) means that radiation inside the star is close to that of a black body at the local temperature. The absence of global equilibrium allows the radiation to leak out through the surface of the star and into space. Thus, stars are objects which attain hydrostatic equilibrium and radiate as near-black bodies for most of their lives.

The early Universe, just after the Big Bang, was much like the interior of the star. The radiation from that epoch has the characteristics of black-body radiation and we see it nowadays as the 2.7 K cosmic background radiation. The early Universe also experienced a period of thermonuclear reactions when most of the hydrogen and helium was produced. The Universe, of course, is not in hydrostatic equilibrium, which has resulted in a variety of dynamical and physical changes as described later in this book.

The thermodynamics that drive the evolution of the Universe and its constituents, can be described in terms of the interaction of particles with radiation and with themselves. The statistical properties of particles and their grouping into "bosons" and "fermions" are key to understanding the statistical physics that underlie the thermodynamics of the Universe. To that end, this part of the book is dedicated to reviewing and using concepts from statistical mechanics to gain a better understanding of astrophysical objects. The part begins with a discussion of particle distributions which are subsequently used to describe the interior structure of stars, the thermodynamics of the early Universe and the properties of radiation. Specific topics include basic thermodynamics, Planck's Law and black-body radiation, the 3 K background radiation and the Big Bang, nucleosynthesis and reaction equilibria in the early Universe, stellar structure, properties of white dwarfs and the properties of neutron stars and pulsars.

Chapter 4

Overview of statistical mechanics

4.1 Thermodynamics

Many astrophysical bodies are close to thermodynamic equilibrium. A proper understanding of them requires that we understand the basic thermodynamic processes that control their properties. I begin with a review of classical and quantum mechanical statistical mechanics. The immediate goal of the review is to derive and discuss the equations that define Bose–Einstein and Fermi–Dirac statistics, the two distributions that govern the behavior of the particles that make up astrophysical bodies.

Let us consider a simple thermodynamic system consisting of two subsystems in physical contact as shown in Fig. 4.1. Each subsystem is characterized by the thermodynamic variables of energy (E), volume (V) and entropy (S). The system is constrained by the conservation of total energy and volume. Once allowed to interact the total system comes into equilibrium when the total entropy no longer changes.

From the first law of thermodynamics

$$\mathrm{d}E = T\,\mathrm{d}S - P\,\mathrm{d}V \tag{81}$$

where T and P are the temperature and pressure, respectively. Solving for $\mathrm{d}S$, and setting it to zero,

$$\mathrm{d}S = \frac{\mathrm{d}E + P\,\mathrm{d}V}{T} = \frac{\mathrm{d}E_1}{T_1} + \frac{P_1\,\mathrm{d}V_1}{T_1} + \frac{\mathrm{d}E_2}{T_2} + \frac{P_2}{T_2}\,\mathrm{d}V_2$$

$$= \left(\frac{1}{T_1} - \frac{1}{T_2}\right)\mathrm{d}E_1 + \left(\frac{P_1}{T_1} - \frac{P_2}{T_2}\right)\mathrm{d}V_1 = 0. \tag{82}$$

We see that the entropy will stop changing when $T_1 = T_2$ and $P_1 = P_2$. In other words, when the total system has achieved *mechanical* and *thermal equilibrium* the entropy is maximized and stops changing.

$$E = E_1 + E_2 \quad = \text{constant}$$
$$V = V_1 + V_2 \quad = \text{constant}$$
$$S = S_1 + S_2 \quad \longrightarrow dS = 0$$

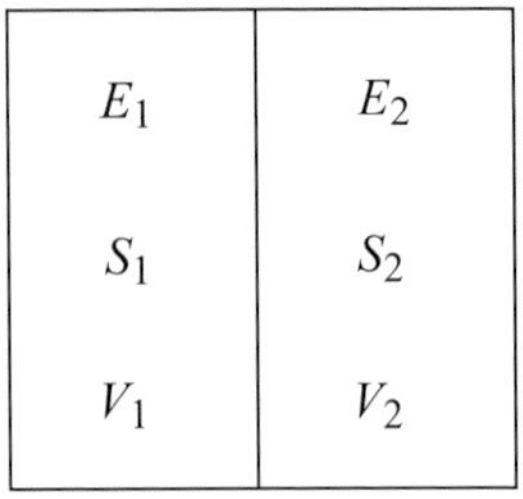

E_1	E_2
S_1	S_2
V_1	V_2

Fig. 4.1 A system consisting of two subsystems. Energy is conserved in the system but not within the subsystems. The two subsystems interact until the entropy stops changing and equilibrium is achieved.

There are other means by which energy can be exchanged. Chemical reactions, for example, can be added to (81), so that

$$dE = T\, dS - P\, dV + \mu\, dN \tag{83}$$

where we have introduced the chemical potential, μ, and the number of particles, N. The result represented by (82) can now be generalized to

$$dS = \left(\frac{1}{T_1} - \frac{1}{T_2}\right) dE_1 + \left(\frac{P_1}{T_1} - \frac{P_2}{T_2}\right) dV_1 - \left(\frac{\mu_1}{T_1} - \frac{\mu_2}{T_2}\right) dN_1 \tag{84}$$

Thus, if $\mu_1 > \mu_2$, $T_1 = T_2$ and $P_1 = P_2 \Rightarrow dN_1 < 0$ for $dS > 0$. With this many variables there are many interrelationships that define how changes in one variable affect other properties. A summary of commonly used thermodynamic terms is as follows:

Energy $\quad E = TS - PV + \mu N \rightarrow dE = T\, dS - P\, dV + \mu\, dN$ $\tag{85}$

Enthalpy $\quad H = E + PV \rightarrow dH = T\, dS + \mu\, dN + V\, dP$ $\tag{86}$

Helmholtz free energy $\quad F = E - TS \rightarrow dF = -S\, dT - P\, dV + \mu\, dN$ $\tag{87}$

Gibbs free energy $\quad G = E - TS + PV \rightarrow dG = -S\, dT + V\, dP + \mu\, dN$ $\tag{88}$

Grand potential $\quad \Omega = E - TS - \mu N \rightarrow d\Omega = -S\, dT - P\, dV - N\, d\mu.$ $\tag{89}$

Combining these equations yields the further relations

$$G = \mu N \qquad \Omega = -PV.$$

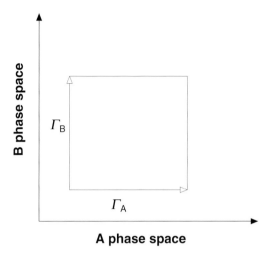

Fig. 4.2 The phase space AB. The individual phase spaces of A and B form a combined phase space AB.

4.2 Classical statistical mechanics

Some basic concepts are now discussed. These are the grand canonical ensemble, the partition function and the grand potential.

A *grand canonical ensemble* is an ensemble of a large number of systems. Each system consists of a subsystem A surrounded by a much larger subsystem B, which together form a closed system containing a large number of particles. Within the subsystem A, the quantities N and E are not necessarily conserved. Consider the schematic of the combined phase space of the system AB, as shown in Fig. 4.2. An elemental volume in the phase space of A and B is given by

$$d\Gamma_A = d^3 q_A \, d^3 p_A$$
$$d\Gamma_B = d^3 q_B \, d^3 p_B.$$

Now, let D_{A+B} be the density of system points in Γ space. The probability of a specific macrostate is given by

$$D_{A+B} = D_A D_B \rightarrow \ln D_{A+B} = \ln D_A + \ln D_B$$

$\Rightarrow D_{A+B} =$ constant. Density D_{A+B} is a conserved quantity! In other words, the system will eventually reach equilibrium, that is $\ln D_{A+B}$ approaches a constant in time.

From our previous discussion regarding conservation laws and the fact that this is a constant sum that looks like a conservation law, we should be able to express $\ln D$ as a linear combination of energy and particle number (since those are the quantities that must be conserved in the system AB) such that

$$\ln D_A = \alpha_A + \beta \mu N_A - \beta E_A.$$

Thus, for subsystem A

$$D = \text{constant} \times e^{\beta(\mu \mathcal{N} - H_\mathcal{N})}$$

where $H_\mathcal{N} = H_\mathcal{N}(q, p)$ is the Hamiltonian representing the total energy of all N_A particles and we have generalized the notation by replacing N_A with $\mathcal{N}$. Dimensionally, β must also have units of energy. Identifying β as the inverse of the thermal energy we have

$$D = \text{constant} \times e^{(\mu \mathcal{N} - H_\mathcal{N})/kT}.$$

This expression requires normalization which can be achieved through the definition

$$D_\mathcal{N} \equiv \frac{1}{Q} e^{(\mu \mathcal{N} - H_\mathcal{N})/kT}. \tag{90}$$

The normalization condition then becomes

$$\sum_{\mathcal{N}=0}^{\infty} \int D_\mathcal{N} \, d\Gamma_\mathcal{N} = \frac{1}{Q} \sum_{\mathcal{N}=0}^{\infty} e^{\mu \mathcal{N}/kT} \int e^{-H_\mathcal{N}(q,p)/kT} \, d^{3\mathcal{N}} q \, d^{3\mathcal{N}} p = 1.$$

Solving this for Q yields

$$Q(T, V, \mu) \equiv \sum_{\mathcal{N}=0}^{\infty} \int e^{(\mu \mathcal{N} - H_\mathcal{N}(q,p))/kT} \, d^{3\mathcal{N}} q \, d^{3\mathcal{N}} p. \tag{91}$$

The integral has the name *canonical partition function* while the summed quantity $Q(T, V, \mu)$ is called the *grand canonical partition function*. We can make the connection with the thermodynamic equations by setting

$$S = -k \sum_{\mathcal{N}=0}^{\infty} \int D_\mathcal{N} \ln D_\mathcal{N} \, d\Gamma_\mathcal{N}. \tag{92}$$

We can see that this is the correct choice by substituting (90) into (92) so that

$$S = -k \sum_{\mathcal{N}=0}^{\infty} \int D_\mathcal{N} \left[\frac{1}{kT} (\mu \mathcal{N} - H_\mathcal{N}) - \ln Q \right] d\Gamma_\mathcal{N}$$

$$= -\frac{\mu}{T} \langle N \rangle + \frac{1}{T} \langle H_\mathcal{N} \rangle + k \ln Q \tag{93}$$

where the angle brackets represent the operation

$$\langle X \rangle \equiv \sum_{\mathcal{N}=0}^{\infty} \int X D_\mathcal{N} \, d\Gamma_\mathcal{N}$$

known as the *grand canonical ensemble average*. Rearrangement of (93) gives

$$TS = -\mu N + E + kT \ln Q. \tag{94}$$

Comparing with (89) yields the definition of Q

$$\Omega \equiv -kT \ln[Q(T, V, \mu)] \tag{95}$$

the grand potential. Thus, the connection with the thermodynamic equations is complete.

4.3 Quantum statistical mechanics

The transition to quantum mechanics requires that we recognize the discreteness of energy levels associated with individual particles.

$$H_\mathcal{N} = \sum_j \mathcal{N}_j \epsilon_j \tag{96}$$

where $\sum_j \mathcal{N}_j = \mathcal{N}$, the total occupation number. The transformation from classical to quantum statistical mechanics therefore requires

$$Q = \sum_{\mathcal{N}=0}^{\infty} \int e^{(\mu\mathcal{N} - H_\mathcal{N})/kT} \, d\Gamma_\mathcal{N} \Rightarrow \sum_{\{\mathcal{N}_j\}} e^{\sum_j \mathcal{N}_j(\mu-\epsilon_j)/kT}. \tag{97}$$

If we use the identity

$$e^{\sum_j x_j} = \prod_j e^{x_j}$$

(97) becomes

$$Q(T, V, \mu) = \sum_{\{\mathcal{N}_j\}} \left[\prod_j y_j^{\mathcal{N}_j} \right] \tag{98}$$

where

$$y_j \equiv e^{(\mu-\epsilon_j)/kT}. \tag{99}$$

If we now use the additional identity

$$\sum_{\{\mathcal{N}_j\}} \left[\prod_j \alpha_j(\mathcal{N}_j) \right] \equiv \prod_j \left[\sum_{\{\mathcal{N}_j\}} \alpha_j(\mathcal{N}_j) \right]$$

we end up with

$$Q(T, V, \mu) = \prod_j \sum_{\mathcal{N}_j} y_j^{\mathcal{N}_j}. \tag{100}$$

This result places us in a better position to discuss Bose–Einstein and Fermi–Dirac statistics.

4.3.1 Bose–Einstein statistics

Recall that power series can have convergent sums, so that

$$\sum_{\mathcal{N}_j=0}^{\infty} y_j^{\mathcal{N}_j} = 1 + y_j + y_j^2 + \cdots = \frac{1}{1 - y_j} \tag{101}$$

where $y < 1$. Combining (100) and (101)

$$\Rightarrow Q = \prod_j \frac{1}{1 - y_j}.$$

Substituting into the definition of the grand potential and using (99) we obtain

$$\Omega = -kT \ln Q = kT \sum_j \ln\left[1 - e^{(\mu-\epsilon_j)/kT}\right]. \tag{102}$$

From (89), $N = -(\partial\Omega/\partial\mu)_{T,V}$ so that

$$N = \sum_j \frac{e^{(\mu-\epsilon_j)/kT}}{1 - e^{(\mu-\epsilon_j)/kT}} = \sum_j \frac{1}{e^{(\epsilon_j-\mu)/kT} - 1}. \tag{103}$$

Identifying the quantity inside the summation sign as the average occupation number, we finally have

$$\langle \mathcal{N}_j \rangle = \frac{1}{e^{(\epsilon_j-\mu)/kT} - 1}. \tag{104}$$

4.3.2 Fermi–Dirac statistics

The Pauli exclusion principle requires that $\mathcal{N}_j = 0, 1$ only. Thus

$$\sum_j y_j^{\mathcal{N}_j} = 1 + y_j$$

so that

$$\Omega = -kT \ln Q = -kT \sum_j \ln\left[1 + e^{(\mu-\epsilon_j)/kT}\right]$$

with

$$N = -\left(\frac{\partial\Omega}{\partial\mu}\right)_{T,V} \equiv \sum_j \langle \mathcal{N}_j \rangle$$

$$\Rightarrow \langle \mathcal{N}_j \rangle = \frac{1}{e^{(\epsilon_j-\mu)/kT} + 1}. \tag{105}$$

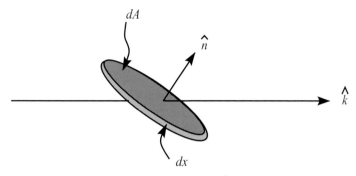

Fig. 4.3 Elemental surface. The beam direction, $\hat{k}$, and the normal to the surface, $\hat{n}$, are shown.

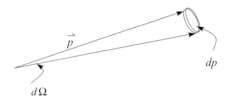

Fig. 4.4 Elemental volume. The relationship between the elemental momentum volume and the elemental solid angle is shown.

We can see that the distribution functions, equations (104) and (105), reflect the different constraints on the occupation number.

4.4   Photon distribution function

Let us define a function that describes the number of photons of spin state (polarization) α, at time t, in a phase element centered at $\vec{x}$, $\vec{p}$. For photons

$$\alpha = 1, 2 \qquad \vec{p} = \hbar\vec{k} = \frac{h\nu}{c}\hat{k}. \tag{106}$$

Thus the radiant energy per phase element is

$$\mathrm{d}E = \sum_{\alpha=1}^{2} h\nu \; F_\alpha(\vec{x}, \vec{p}, t) \, \mathrm{d}^3x \, \mathrm{d}^3p. \tag{107}$$

For a beam of photons (Figs. 4.3 and 4.4)

$$\mathrm{d}V = \mathrm{d}^3x = \mathrm{d}x \mathrm{d}A = \hat{k} \cdot \hat{n}c \, \mathrm{d}t \, \mathrm{d}A \tag{108}$$

$$\mathrm{d}^3p = p^2 \, \mathrm{d}\Omega \, \mathrm{d}p = \frac{h^3\nu^2}{c^3} \, \mathrm{d}\Omega \, \mathrm{d}\nu \qquad \text{from} \quad (106). \tag{109}$$

Combining (107), (108) and (109)

$$\Rightarrow dE = \sum_{1}^{2} h\nu F_\alpha(\vec{x}, \vec{p}, t)\hat{k} \cdot \hat{n}\, c\, dt\, dA\, \frac{h^3\nu^2}{c^3}\, d\Omega\, d\nu$$

$$= \frac{h^4\nu^3}{c^2} \sum_{1}^{2} F_\alpha(\vec{x}, \vec{p}, t)\hat{k} \cdot \hat{n}\, dA\, d\Omega\, d\nu\, dt. \qquad (110)$$

Let us now compare (110) with the definition of specific intensity, $I_\nu(\hat{k}, \vec{x}, t)$,

$$dE = I_\nu(\hat{k}, \vec{x}, t)\hat{k} \cdot \hat{n}\, dA\, d\Omega\, d\nu\, dt.$$

We see that

$$I_\nu = \sum_{1}^{2} \frac{h^4\nu^3}{c^2}\, F_\alpha(\vec{x}, \vec{p}, t). \qquad (111)$$

This places us in a position to define an occupation number for each spin state, that is an occupation number per *phase cell*

$$\mathcal{N}_\alpha \equiv h^3 F_\alpha \qquad (112)$$

so that (111) can be rewritten as

$$I_\nu = \sum_{1}^{2} \frac{h\nu^3}{c^2}\mathcal{N}_\alpha(\vec{x}, \vec{p}, t). \qquad (113)$$

Equation (113) represents a relationship between the intensity of a beam of photons and the occupation number which describes the statistical behavior of the photons with respect to the phase space. Under thermodynamic equilibrium, (111) and (113) are constrained to specific functional forms. We now consider those.

4.5   Thermodynamic equilibrium

Kirchhoff showed that under thermodynamic equilibrium

$$I_\nu = B_\nu(T) \qquad (114)$$

the spectrum of emitted radiation depends only on the temperature of the emitting body. Planck showed that the specific functional form of B_ν is given by

$$B_\nu = \frac{2h\nu^3/c^2}{e^{h\nu/kT} - 1}. \qquad (115)$$

Proper derivation of (115) from (113) requires knowledge of Bose–Einstein statistics given that photons are bosons. Since we defined the latter in the previous

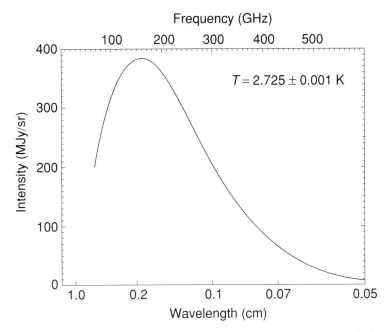

Fig. 4.5 The 3 K background spectrum. The measurements were made by the FIRAS instrument on COBE. The best curve yields a temperature of 2.725 K. This figure was obtained from http://space.gsfc.nasa.gov/astro/cobe.

subsection let us do that. Using (104) and the knowledge that for photons, $\mu = 0$ we have

$$\mathcal{N}_\alpha = \frac{1}{e^{h\nu/kT} - 1}.$$ (116)

Substituting (116) into (113) yields (114) and (115). It is useful to examine these results in limits of high and low photon energy.

For $h\nu \gg kT$

$$\mathcal{N}_\alpha = e^{-h\nu/kT}$$

which represents the classical *Boltzmann distribution*. Photons act like particles.

For $h\nu \ll kT$

$$\mathcal{N}_\alpha = \frac{kT}{h\nu} \gg 1$$

which is in the *Rayleigh–Jeans Limit*. In a single phase cell, $E = h\nu \mathcal{N}_\alpha = kT$ and photons act like waves.

An example of a black-body spectrum is shown in Fig. 4.5. It is the spectrum expected from a body in thermodynamic equilibrium at a temperature of 2.7 K. It is also the best fit curve for measurements made of the cosmic background radiation with the FIRAS detector on the COBE satellite (see COBE homepage – http://space.gsfc.nasa.gov/astro/cobe). The curve in Fig. 4.5 represents a perfect black body to within the experimental errors, which tells us that the cosmic 3 K radiation was originally emitted under conditions of thermodynamic equilibrium. Since the modern day Universe consists of both matter and radiation, these two components were, at one time, in equilibrium. When was that and what is the implication of this result on galaxy formation? That is the subject of the next chapter.

4.6 Further reading

Garrod, C. (1994) *Statistical Mechanics and Thermodynamics*, Oxford University Press, New York, NY, USA.

Kittel, C. (1980) *Thermal Physics*, 2nd edn, Freeman, San Francisco, CA, USA.

Pauli, W. (2000) *Thermodynamics and the Kinetic Theory of Gases*, Dover, New York, NY, USA.

Shu, F. H. (1992) *The Physics of Astrophysics: Radiation*, University Science Books, Mill Valley, CA, USA.

Chapter 5

The early Universe

5.1 The 3 K background radiation

According to the COBE results, the thermal properties of the cosmic background radiation are characterized by a temperature of

$$T = 2.725 \pm 0.006 \, \text{K}$$

and temperature fluctuations of

$$\frac{\Delta T}{T} = 6 \times 10^{-6}$$

which represent the spatial variation of temperature across the sky. These results have fundamental importance with respect to our knowledge of the early Universe and the formation of galaxies. Given our earlier discussion of the gravitational dynamics of the Universe and given what we now know about the background radiation, let us evolve the "radiation" Universe back in time to see what we can learn.

5.1.1 History of the background radiation

According to energy conservation and the equations developed in Section 3.1

$$E_T = E_k + E_G = \frac{2\pi}{5}\rho H^2 R^5 - \frac{16\pi^2}{15}\rho^2 G R^5 = \text{constant.} \qquad (117)$$

We know that $\rho = M/(\frac{4}{3}\pi R^3)$, $H = v/R$ so that (117) can be cast into the form

$$E_T = 0.3Mv^2 - \frac{3}{5}GM^2R^{-1}$$

$$\Rightarrow v^2 - \frac{2GM}{R} = \frac{E_T}{M}\frac{10}{3} = \mathcal{E}$$

$$\Rightarrow \frac{\mathrm{d}R}{\mathrm{d}t} = \sqrt{\mathcal{E} + \frac{2GM}{R}} \qquad (118)$$

69

$$\Rightarrow \frac{dR}{\sqrt{\mathcal{E} + (2GM/R)}} = dt$$

$$\Rightarrow \int_0^R \frac{dR'}{\sqrt{\mathcal{E} + (2GM/R')}} = \int_0^t dt'.$$

This equation represents the evolution of $R(t)$ for a given energy and mass of the Universe.

We can consider the three special cases corresponding to *flat*, *open* and *closed* Universes.

Flat Universe ($\mathcal{E} = 0$)

In this case, the integral is trivial

$$\frac{1}{\sqrt{2GM}} \int_0^R dR' \sqrt{R'} = t$$

$$\Rightarrow R(t) = \left(\frac{3}{2}\right)^{2/3} (2GM)^{1/3} t^{2/3} = \left(\frac{9}{2} GM\right)^{1/3} t^{2/3}. \qquad (119)$$

The radius increases as $t^{2/3}$.

Open Universe ($\mathcal{E} > 0$)

When the Universe has a positive total energy its long-term evolution is determined by the excess kinetic energy. In that case, the solution to (118) is more complex. The author has used the software package MAPLE to obtain the solution below.

$$\sqrt{\frac{R^2}{\mathcal{E}} + \frac{2GMR}{\mathcal{E}^2}} - \frac{GM}{\mathcal{E}^{3/2}} \ln\left[2\sqrt{\mathcal{E}^2 R_2^2 \mathcal{E} GMR + 2\mathcal{E}R + 2GM}\right] = t. \qquad (120)$$

Closed Universe ($\mathcal{E} < 0$)

When the total energy is negative the gravitational attraction of the matter wins out in the long term and the Universe eventually collapses. The solution, obtained in a similar way, becomes

$$\Rightarrow \frac{GM}{\mathcal{E}^{3/2}} \arcsin\left[\frac{\mathcal{E}R}{GM} - 1\right] - \sqrt{\frac{2GMR}{\mathcal{E}^2} - \frac{R^2}{\mathcal{E}}} = t. \qquad (121)$$

5.1.2 Evolution of energy density

Recall that the quantity $\mathcal{E}$ represents energy density. Let us now consider the energy density of the photons.

$$\mathcal{E}_\gamma = n_\gamma h\nu. \qquad (122)$$

But the number density, N_γ has the behavior $n_\gamma \propto R^{-3}$ and the frequency of a photon goes as $\nu \propto 1/\lambda \propto 1/R$ as a result of being redshifted by the expansion. Thus, according to (122)

$$\mathcal{E}_\gamma \propto R^{-4}.$$

But, $\mathcal{E}_\gamma \propto T^4 \Rightarrow T \propto 1/R$. Combining all of this, we have

$$\Rightarrow \frac{T}{T_0} = \frac{R_0}{R}. \tag{123}$$

Combining (123) and (119)

$$\Rightarrow \frac{T}{T_0} = \left(\frac{t_0}{t}\right)^{2/3} \tag{124}$$

for a flat Universe. This is a good scaling law even for $\mathcal{E} \neq 0$ so long as $\mathcal{E}$ is close to 0. The scaling law is valid only for $t > t_D$ after the decoupling time. For times before decoupling the radiation pressure of the photons dominated the dynamics of the Universe.

It is now possible to evolve the Universe back in time using this scaling law. As starting points we will use contemporary values of $T_0 = 2.7$ K and $t_0 = 1.4 \times 10^{10}$ years. Let us evolve the Universe back in time when $T = 3000$ K, the era of decoupling. The age of the Universe would then have been

$$t_D = \left(\frac{T_0}{T}\right)^{3/2} \qquad t_0 = 4 \times 10^5 \text{ years.}$$

Thus, for times $t \leq 4 \times 10^5$ years, radiation and matter would have been tightly coupled because the photons were energetic enough to keep matter ionized. During this time the Universe was radiation dominated in the sense that radiation energy density was greater than the energy density of matter. Any clustering of matter during this epoch would have to be accompanied by a clustering of radiation energy density (i.e. a temperature enhancement).

5.2 Galaxy formation

In the hot-dark-matter (HDM) top-down scenario, the modern day structure requires that

$$\frac{\delta\rho}{\rho} > 10^{-4}$$

at time of decoupling, where ρ is the matter density. Otherwise Hubble expansion and gravitation alone could not produce the modern day large scale structure. Just

before decoupling, matter and radiation were coupled so that the density enhancements must also show up as temperature enhancements

$$\frac{\delta\rho}{\rho} = \frac{\delta(T^4)}{T^4} = \frac{4\delta T}{T} \tag{125}$$

so that the HDM model would predict

$$\frac{\delta\rho}{\rho} > 10^{-4} \Rightarrow \frac{\delta T}{T} > 2.5 \times 10^{-5}.$$

Now, according to COBE, $\delta T / T \approx 6 \times 10^{-6}$ K, which argues against a purely
HDM model.

Now, some cold-dark-matter (CDM) models predict

$$\frac{\delta T}{T} > 4 \times 10^{-6} \text{ K}$$

and they cannot be ruled out by the COBE observations. On the other hand, CDM
has difficulty accounting for the largest structures in the Universe. For these reasons
hybrid models, combining HDM and CDM, were introduced. However, the possibility of a nonzero cosmological constant makes the hybrid models less relevant.
In fact, CDM, in conjunction with a nonzero Λ can satisfactorily account for the
large scale structure, as discussed in Chapter 3.

5.3   Local cosmology and nucleosynthesis

5.3.1   Overview

As noted in the previous section, the energy density of the Universe was much higher
in the past. Before decoupling, matter and radiation were in close equilibrium and
there were times during the early evolution of the Universe when the conditions were
similar to those found in the interiors of stars such as the Sun. In other words, the
density, temperature and pressure were high enough for thermonuclear reactions
to take place and for nucleosynthesis to occur. Consequently the following are
examples of cosmological questions.

- Why is 10^{-5} of the seawater "heavy"? In other words why is the ratio [D/H] $\approx 10^{-5}$?
- Why is 25% of the Sun's mass in the form of ^{4}He?

The reason the above are cosmological questions is that there is no known process in
the current Universe to account for the above ratios. There is simply too much ^{2}D and
^{4}He (and ^{7}Li) around. Yet we know that these elements can be produced in simple
nuclear reactions whose physics are understood. With scaling laws in hand, we know
that if we evolve the Universe back in time the conditions for nucleosynthesis existed

for $t \leq 1$ ms. Having estimated these conditions and knowing how nuclear reactions operate (they are governed by readily understood laws of physics) it is possible for us to estimate how much ^{2}D and ^{4}He (and ^{7}Li) could have been produced during the Big Bang. Such calculations indicate that the currently observed abundances of these elements can be entirely explained as resulting from the Big Bang. The success of these calculations gives great weight to the Big Bang hypothesis and gives researchers the confidence to examine the physics of the Universe back to $t \approx 10^{-35}$ seconds and to contemplate the birth of the Universe and its basic elements.

Despite the fact that BBN can account for the current abundances of ^{2}D and ^{4}He (and ^{7}Li) we should still ask the question – can these elements be produced in a more conventional manner? Could ^{2}D and ^{4}He (and ^{7}Li) be products of stellar nucleosynthesis (such as takes place in the core of the Sun)?

5.3.2 Primordial helium

All main-sequence stars produce ^{4}He in their cores. These elements are dispersed into the interstellar medium through stellar winds, ejection of planetary nebulas and supernova explosions. So why cannot these processes account for the current abundance of ^{4}He in the Universe? Let us answer that question quantitatively by examining how much energy is needed to make a ^{4}He atom from 4 nucleons. We start off with the definition of an atomic mass unit (amu),

$$1 \text{ amu} = 931.478 \text{ MeV}/c^2.$$

Table 5.1 lists the mass defects of some simple elements.

When $4p \rightarrow {}^4\text{He} + Q$, the energy released is $Q = 26.7$ MeV, of which about 1.6 MeV is in the form of neutrinos. Thus, the formation of the ^{4}He atom is accompanied by the release of > 25 MeV $\approx 4 \times 10^{-5}$ ergs.

If we now take all the nucleons in the Universe

$$\langle n_{\text{nucleons}} \rangle \geq 10^{-7} \text{ cm}^{-3}$$

Table 5.1. *Mass defects*

Element	$(M - A) \times$ amu c^2	$(M - A)/A$
n	8.07 MeV	8.07 MeV
p	7.29 MeV	7.29 MeV
^{2}H = D	13.14 MeV	6.57 MeV
^{3}He	14.93 MeV	4.98 MeV
^{4}He = α	2.42 MeV	0.61 MeV
^{12}C	0	0

we can make

$$\rightarrow \langle n_{(^4\mathrm{He})} \rangle \approx 0.25 \frac{\mathrm{nucleon}}{4} \geq 6 \times 10^{-9} \ \mathrm{cm}^{-3}.$$

helium nuclei. The energy associated with making this many helium nuclei is then given by

$$\langle \mathcal{E}_\alpha \rangle \approx 4 \times 10^{-5} \ \mathrm{erg} \times 6 \times 10^{-9} \ \mathrm{cm}^{-3} = 2 \times 10^{-13} \ \mathrm{erg} \ \mathrm{cm}^{-3}.$$

Now, most of the luminosity of the Universe arises ultimately from the production of ^{4}He. The mean luminosity density of the Universe is known to be $\approx 2 \times 10^{-32} \ \mathrm{erg \ s}^{-1} \ \mathrm{cm}^{-3}$. Thus, in the 10^{10} years $= 3 \times 10^{17}$ seconds, during which stars have existed, the amount of energy arising from ^{4}He production is

$$\langle \mathcal{E}_L \rangle \approx 6 \times 10^{-15} \ \mathrm{erg} \ \mathrm{cm}^{-3}.$$

Thus, the ratio of energy densities resulting from actual helium production to that needed to account for the observed amount is

$$\frac{\langle \mathcal{E}_L \rangle}{\langle \mathcal{E}_\alpha \rangle} \approx 0.03.$$

Thus, if *all* the luminosity of the Universe were used to make ^{4}He over the entire age of the Universe it could change the present amount by only a few percent. Thus the ^{4}He must be *primordial*. What about D?

The number density of D is

$$\langle n_D \rangle = 10^{-5} \langle n_N \rangle \approx 10^{-12} \ \mathrm{cm}^{-3}.$$

The element D is most easily made in the reaction

$$\mathrm{n} + \mathrm{p} \rightarrow \mathrm{D} + Q; \qquad Q = 2.2 \ \mathrm{MeV}$$
$$\mathrm{p} + \mathrm{p} \rightarrow Q + Q; \qquad Q = 3 \ \mathrm{MeV}$$

so that the energy density associated with the formation of D is given by

$$\langle \mathcal{E}_D \rangle \approx 3 \ \mathrm{MeV} \times 10^{-12} \ \mathrm{cm}^{-3} \approx 5 \times 10^{-18} \ \mathrm{erg \ cm}^{-3} \ll \langle \mathcal{E}_L \rangle.$$

There is no problem accounting for D given the current nuclear reactions going on in the Universe. However, D is very hard to preserve once it is made. The environments in which it is currently made (stellar interiors) are conducive to the destruction of D. This can best be illustrated in the context of *nuclear reaction rates* which we discuss now.

5.4 Reaction rates

5.4.1 Introduction

Consider the generic reaction

$$a + X \rightarrow Y + b + Q$$

where X is stationary and there is a flux of incident particles a having a density n_a and velocity v. Then

$$\text{flux} = n_a v$$

$$\text{cross-section} = \sigma(v) \equiv \frac{\text{reactions/time (per X particle)}}{\text{flux of } a \text{ (particles area}^{-1} \text{ time}^{-1})}.$$

So that the number of reactions volume^{-1} second^{-1} are

$$r_{aX} = n_a n_X \langle \sigma v \rangle$$

where v is the relative velocity and

$$\langle \sigma v \rangle \equiv \int_0^{\infty} \sigma(v)\phi(v) v \, dv$$

where $\phi(v)$ is the normalized relative velocity distribution such that

$$\int \phi(v) \, dv = 1.$$

For a non-degenerate gas (defined in Section 5.5)

$$\phi(v) \, dv \propto \underbrace{e^{-E/kT}}_{\text{Boltzmann factor}} \quad \underbrace{v^2 \, dv}_{\text{phase space factor}}$$

$$\phi(v) = \left(\frac{\mu}{2\pi kT}\right)^{3/2} e^{[-(\mu v^2)/(2kT)]} 4\pi v^2$$

$$\mu = \frac{m_a m_X}{m_a + m_X} = \text{reduced mass.}$$

The reaction cross-section is governed by nuclear attraction at small distances and Coulomb repulsion at larger distances. The nuclear attraction represents a binding energy which is typically around 8 MeV/nucleon (see Table 5.1). Thus, Q_{reaction} is always $\approx$ MeV.

The range of the attractive nuclear force is given by

$$R \approx A^{1/3} \times 10^{-13} \text{ cm} = A^{1/3} \text{ fm.}$$

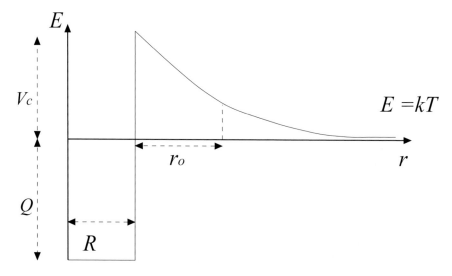

Fig. 5.1 The Coulomb barrier. A particle with $E = kT$ traveling leftward from $r = \infty$ will see a Coulomb barrier of amplitude V_c and scale r_0. The nuclear potential well, depth Q and scale R is also shown.

On the other hand, Coulomb repulsion is described by

$$V = \frac{q_a q_X}{r} = Z_a Z_X \times \frac{e^2}{R}$$
$$= 1.44 \frac{Z_a Z_X}{R(\text{fm})} \text{ MeV}.$$

The excess energy is the result of having to overcome the Coulomb barrier. Typically, the kinetic energies of the particles undergoing nuclear reactions have

$$E_k \approx kT \approx 1 \text{ keV} \quad (\text{Sun})$$
$$E_k \approx kT \approx 10 \text{ keV} \quad (\text{BBN}).$$

We see that $E_k \ll V_{coulomb}$ so that *classically* nuclear reactions are impossible. The distance r_0 at which the Coulomb potential equals the kinetic energy of the incident particle (Fig. 5.1) is then $r_0 \approx 10^3 R$. Classically the incident particle turns back very far from the region where nuclear reactions can take place. The reason that nuclear reactions take place at all is because of quantum mechanical tunneling. We now review the concept of tunneling by discussing the QM 1-D potential barrier.

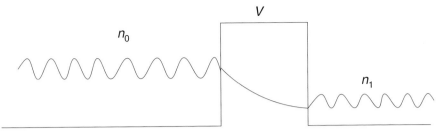

Fig. 5.2 Quantum tunneling. A particle propagating from left to right has a finite chance of tunneling through the potential barrier V.

5.4.2 Barrier penetration

We can represent the concept of Coulomb barrier penetration by considering Fig. 5.2. Recall some basic quantum mechanical terms like *wave function*, Ψ and the *probability density*, $n \propto \Psi^2$. Writing down Schrödinger's equation

$$E\Psi = H\Psi = \left(\frac{p^2}{2m} + V \right) \Psi$$

where E is the particle energy, V is the barrier potential and p is the particle momentum. Recall that the momentum can be thought of as an operator such that

$$p \equiv -i\hbar \frac{\partial}{\partial x}$$

in one spatial dimension. Thus, the Schrödinger equation can be recast into the form

$$E\Psi = -\frac{\hbar^2}{2m} \frac{\partial^2}{\partial x^2} \Psi + V\Psi$$

or

$$\frac{\partial^2}{\partial x^2} \Psi = \frac{2m}{\hbar^2} (V - E)\Psi.$$

If there were no barrier, then

$$E = \frac{mv^2}{2} \qquad V = 0$$

and

$$\frac{\partial^2}{\partial x^2} \Psi = -\frac{m^2 v^2}{\hbar^2} \Psi = -k^2 \Psi.$$

This is just the equation for a traveling wave, meaning that the particle travels along unimpeded. The wave function then becomes

$$\Psi = \Psi_0 \, e^{\pm ikx}$$

and particle motion is possible in either direction. The probability density is then given by

$$|\Psi|^2 = \Psi \Psi^* = |\Psi_0|^2 = \text{constant}$$

as expected.

With the barrier in place, $V > 0$ and we have

$$\frac{\partial^2}{\partial x^2}\Psi = \frac{2m}{\hbar^2}(V - E)\Psi \equiv \lambda^2 \Psi$$

for which the solution is

$$\Psi = \Psi_0 \, e^{-\lambda x} + \underbrace{\Psi_1 \, e^{+\lambda x}}_{\to 0}$$

since we require that the probability density be bounded over an infinite volume.

Thus, the probability density is given by

$$|\Psi|^2 = |\Psi_0|^2 \, e^{-2\lambda x}.$$

The penetration is characterized by the parameter

$$P = \frac{|\Psi(x_b)|^2}{|\Psi_0|^2} = e^{-2\lambda x_b}.$$

where, $x_b = 1/(2\lambda)$ is the scale length for the penetration and

$$\lambda = \left[\frac{2m}{\hbar^2}(V - E)\right]^{1/2}.$$

The larger the V and the smaller the energy of the particle the harder it is for barrier penetration to work. However, the exponential dependence of the effect ensures that the probability for penetration is always finite.

Gamow Peak

In real life $V = V(x)$ thereby complicating this simple picture. These results would have to be generalized so that

$$\ln P \approx 2 \int \lambda(x)\,\mathrm{d}x.$$

Consequently, the relevant integral is

$$\ln P = 2 \int_{r_0}^{R} \lambda(r)\, dr$$

$$= 2 \int_{r_0}^{R} \left[\frac{2m}{\hbar^2} \left(\frac{Z_a Z_X e^2}{r} - \frac{m v_0^2}{2} \right) \right]^{1/2} dr.$$

The classical turning point is defined by the condition

$$\frac{Z_a Z_X e^2}{r_0} \equiv \frac{m v_0^2}{2}.$$

Thus

$$\ln P = 2 \left[\left(\frac{2m}{\hbar^2} \right) \left(\frac{m v_0^2}{2} \right) \right]^{1/2} \int_{r_0}^{R} \left(\frac{r_0}{r} - 1 \right)^{1/2} dr$$

$$= 2 \left[\left(\frac{2m}{\hbar^2} \right) \left(\frac{m v_0^2}{2} \right) \right]^{1/2} r_0 \underbrace{\int_{1}^{R/r_0} \left(\frac{1}{x} - 1 \right)^{1/2} dx}_{\text{convergent for } x \to 0}$$

where $x = r/r_0$. Setting the integral to a finite value I, we have

$$\ln P = 2 \left[\left(\frac{2m}{\hbar^2} \right) \left(\frac{m v_0^2}{2} \right) \right]^{1/2} \frac{2 Z_a Z_X e^2}{m v_0^2} I = \frac{4I Z_a Z_X e^2}{\hbar v_0}.$$

The exact solution, generalized for any v, is

$$P = \mathrm{e}^{-(2\pi Z_a Z_X e^2)/(\hbar v)}.$$

The convention for defining the cross-section in terms of the penetration parameter is given by the equation

$$\sigma(E) = \frac{S(E)}{E}\, \mathrm{e}^{-(2\pi Z_a Z_X e^2)/(\hbar v)}.$$

The function $S(E)$ depends on the nuclear physics and is weakly dependent on the energy. Armed with this definition of σ let us turn our attention back to reaction rates

$$\langle \sigma v \rangle = \int_0^\infty \sigma \phi v\, dv = \int \sigma \phi \frac{dE}{\mu}$$

where

$$\sigma(E) = \frac{S(E)}{E}\, \mathrm{e}^{-b/\sqrt{E}}$$

$$\phi = \left(\frac{\mu}{2\pi kT} \right)^{3/2} \mathrm{e}^{-E/kT} \left(\frac{8\pi}{\mu} \right) E$$

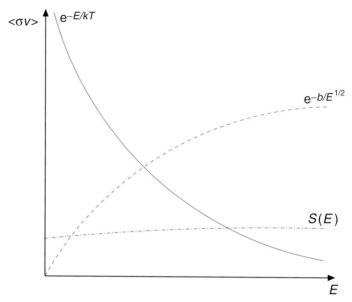

Fig. 5.3 Components of the reaction rate function. The shapes of the three functions that make up the expression for $\langle \sigma v \rangle$ are shown.

$$b \equiv \frac{2\pi Z_a Z_X e^2}{\hbar} \left(\frac{\mu}{2}\right)^{1/2}$$

$$\langle \sigma v \rangle = \left(\frac{8}{\mu \pi}\right)^{1/2} \frac{1}{(kT)^{3/2}} \int_0^\infty S(E) \, e^{[-(E/kT)-(b/\sqrt{E})]} \, dE.$$

Figure 5.3 illustrates the dependence of $\langle \sigma v \rangle$ on energy. The exponential in the integral is sharply peaked at some energy E_0. Let us find out what E_0 is. We begin by searching for a local extremum in the argument of the exponential

$$\frac{d}{dE}\left[\frac{E}{kT} + \frac{b}{\sqrt{E}}\right] = 0 \qquad \text{at } E_0$$

$$\frac{1}{kT} - \frac{b}{2E_0^{3/2}} = 0$$

$$\Rightarrow E_0 = \left(\frac{bkT}{2}\right)^{2/3}$$

$$E_0 = 1.2 \left(Z_a^2 Z_X^2 A_{red} T_6^2\right)^{1/3} \text{ keV}.$$

This represents the optimal energy for nuclear reactions. The sharp drop off in energy on either side of E_0 gives the cross-section a peaked geometry with respect to energy. The region centered on E_0 is known as the *Gamow Peak*.

5.4.3 Estimating reaction rates

All that remains now is to estimate the integral in order to obtain a numerical expression for the reaction rate. Integrals of the form

$$I = \int_A^B S(E) e^{-f(E)} \, dE$$

where $f(E)$ is peaked at E_0 can be estimated by the *method of steepest descent*. Part of the method involves an expansion of $f(E)$ about E_0 in a Taylor series.

$$f(E) \approx f(E_0) + \frac{\partial f}{\partial E}(E - E_0) + \left(\frac{\partial^2 f}{\partial E^2}\right)_{E_0} \frac{(E - E_0)^2}{2} + \cdots$$

$$\approx f(E_0) + \left(\frac{\partial^2 f}{\partial E^2}\right)_{E_0} \frac{(E - E_0)^2}{2}$$

$$= f(E_0) + \left(\frac{E - E_0}{\sigma}\right)^2$$

where

$$\sigma^2 = \frac{2}{(\partial^2 f/\partial E^2)_{E_0}}$$

$$I \approx e^{-f(E_0)} S(E_0) \int_{-\infty}^{\infty} e^{-[(E-E_0)/\sigma]^2} \, dE$$

$$= \sigma \sqrt{\pi} S(E_0) e^{-f(E_0)}.$$

In our case

$$f(E) = \frac{E}{kT} + \frac{b}{\sqrt{E}}$$

$$E_0 = \left(\frac{bkT}{2}\right)^{2/3} \gg kT$$

$$f(E_0) = 3E_0$$

$$\sigma = \frac{2}{\sqrt{3}}(E_0 kT)^{1/2} \ll E_0$$

$$\Rightarrow r_{aX} = n_a n_X \frac{7 \times 10^{-19}}{A_{red} Z_a Z_X} S_0 (\text{keV barns}) \tau^2 e^{-\tau} \, s^{-1} \, cm^{-3}$$

where

$$\tau = 42.5 \left(\frac{Z_a^2 Z_X^2 A_{red}}{T_6}\right)^{1/3}$$

where

$$T_6 = T/10^6 \text{ K} \qquad A_{red} = \left(\frac{1}{A_a} + \frac{1}{A_X} \right)^{-1} = \frac{A_a A_X}{A_a + A_X}$$

Note the very strong dependence on $Z_a Z_X$. The factor of 42.5 also gives the reaction rate a strong dependence on T_6.

5.4.4 Destruction of D

It turns out that

$$D + p \rightarrow {}^3\text{He} + \gamma$$

is the most rapid nuclear reaction. If we examine the deuterium lifetime

$$\frac{1}{t_D} = \frac{1}{n_D} \left(\frac{\partial n_D}{\partial t} \right) = n_p \langle \sigma v \rangle_{D_p}$$

we see that it is short because $Z_D Z_p = 1$ has the lowest possible value and protons are the most abundant constituent. In any environment where D can be made it will be destroyed. In the Sun

$$[\text{D/H}] \approx 10^{-18}$$

reflecting the fact that D is easily destroyed. The destruction of D is what makes it difficult to preserve. Therefore, the observed cosmic abundance of D can only be explained if it is primordial. How much D is produced in BBN is critically dependent on the conditions that existed in the early Universe. Explaining its current abundance tells us much about the nature of the early Universe.

5.4.5 Formation of D

When $T < 10^9$ K the following reaction forms deuterium

$$n + p \rightarrow D + \gamma.$$

However, the D that is formed then participates in a subsequent reaction

$$D + D \rightarrow {}^4\text{He} + \gamma.$$

Whereas, the former reaction produces D, the latter one destroys it. The relative rates of the two reactions determine the equilibrium abundance of the deuterium. Generally, the second reaction is the faster of the two and it therefore determines the equilibrium value of D.

Using the preceding arguments, the timescale for the DD reaction is

$$t_{DD} \propto \rho_b^{-1}.$$

where ρ_b is the baryon mass density. The greater the mass density of the Universe, for example, the faster the DD reaction proceeds and the lower the equilibrium value of D. The lower the mass density the greater the D abundance. It is for this reason that the D abundance is a sensitive tracer of the conditions (density) in the early Universe.

5.4.6 Formation of ^{4}He

Before Big Bang nucleosynthesis began, the early Universe had plenty of free neutrons (see next section) and protons. The interaction of n and p leads to the production of ^{4}He, as we have noted. If the abundance of n is significantly lower than the abundance of p essentially all the n go into making ^{4}He. Thus, the abundance of ^{4}He is given by

$$^4\text{He} \approx \frac{n/2}{(n+p)/4} = \frac{2n}{n+p}.$$

The contemporary ratio of neutrons to protons is about 0.12. Inserting that value into the previous equation yields ^{4}He $\approx 2/9$, which is the approximate abundance of ^{4}He in the Universe today.

5.5 Particle equilibria in the early Universe

5.5.1 Overview

The formation of elements in the early Universe can best be understood by considering the environments in which current fundamental particles were in equilibrium. These particles include well-known baryons such as neutrons and protons, leptons such as electrons, mu neutrinos, electron neutrinos, tau neutrinos and antineutrinos, as well as photons. All the particles are, of course, fermions while the photons are bosons. For the particles to be in equilibrium with each other their kinetic energies must be comparable to their rest mass energies. Table 5.2 lists the rest mass energies of some particles. For temperatures above about 10^{12} K all of the particles are in equilibrium with each other.

Using the temperature scaling law appropriate for cosmic expansion before decoupling

$$t = t_d \times \left(\frac{T_d}{T_{12}}\right)^2$$

is the time after the Big Bang when the temperature was 10^{12} K. Using the numbers from Section 5.1.5, we obtain $t_{12} = 10^{-4}$ seconds. Thus, it appears that 10^{-4} seconds after the Big Bang there was an equilibrium between n, p, e^{+-}, γ, ν_e, $\bar{\nu}_e$, ν_μ, $\bar{\nu}_\mu$, ν_τ, $\bar{\nu}_\tau$.

Table 5.2. *Particle equilibrium temperatures*

Particle species A	Rest mass energy	T_{eq}
Electron/positron	511 keV	6×10^9 K
Muon/antimuon	107 MeV	1.2×10^{12} K
μ and e neutrinos/antineutrinos	0	0
Proton/neutron	940 MeV	10^{13} K*
Pions	140 MeV	1.6×10^{12} K

* $T_n - T_p = 1.5 \times 10^{10}$ K

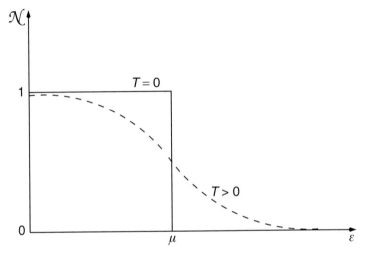

Fig. 5.4 Fermion occupancies at $T = 0$ and $T > 0$. At $T = 0$ all states up to μ are filled. At finite temperatures the sharp boundary at μ is softened and the distribution assumes a more classical form.

The statistical distribution of the fermions is described by (105). Let us examine the behavior of the occupation number in the limits of strongly negative and strongly positive chemical potentials because this behavior will illustrate how fermions in the Universe behave.

For $\mu \ll -kT$ equation (105) reduces to

$$\mathcal{N} \approx e^{(\mu-\epsilon)/kT} = e^{-|\mu/kT|} e^{-\epsilon/kT} = \alpha e^{-\epsilon/kT}$$

where $\alpha \ll 1$. The behavior is exponential, as you would expect.

For $\mu \gg kT$ we have $\mathcal{N} \approx 1$ if $\epsilon \ll \mu$ and $\mathcal{N} \ll 1$ if $\epsilon \gg \mu$. The corresponding distribution function is shown in Fig. 5.4 ($T = 0$).

The intermediate case, $\mu \approx kT$, is an interpolation between the two extreme cases and is shown in Fig. 5.4 ($T > 0$). The number density follows from the occupation number according to

$$n = \sum_\alpha \int \frac{d^3p}{h^3} \mathcal{N}_\alpha(\epsilon). \tag{126}$$

In the low-density, non-degenerate limit $\mu \ll kT$ we have the *Maxwell–Boltzmann* limit

$$n = \sum_\alpha \int \frac{d^3p}{h^3} e^{\mu/kT} e^{-\epsilon/kT} \tag{127}$$

where

$$\epsilon = \underbrace{\epsilon_{kinetic}}_{\epsilon^k = p^2/(2m)} + \underbrace{\epsilon_{internal}}_{\underbrace{\epsilon_0^\alpha}_{\text{lowest level}} + \underbrace{\epsilon_i^\alpha}_{\text{excitations}}}$$

$$n = \sum_\alpha \int_0^\infty \frac{4\pi p^2 \, dp}{h^3} (e^{\mu/kT}) \, e^{-\epsilon^k/kT} e^{-\epsilon_0/kT} e^{-\epsilon_i^\alpha/kT}$$

$$= \frac{4\pi \, e^{\mu/kT}}{h^3} \underbrace{\left(\int p^2 \, dp \, e^{-\epsilon^k/kT} \right)}_{\int p(p \, dp) \, e^{-\epsilon^k/kT}} \underbrace{\left(\sum_\alpha e^{\epsilon_i^\alpha/kT} \right)}_{g \equiv \text{internal partition function}} e^{-\epsilon_0/kT}. \tag{128}$$

For non-relativistic particles

$$p = (2m\epsilon^k)^{1/2} \quad p \, dp = m \, d\epsilon^k$$

so that

$$\int p^2 \, dp \, e^{-\epsilon^k/kT} = \frac{1}{2}(2mkT)^{3/2} \underbrace{\int_0^\infty \sqrt{x} \, e^{-x} \, dx}_{\sqrt{\pi}/2}$$

so that (128) becomes

$$n = e^{\mu/kT} \left(\frac{2\pi mkT}{h^2} \right)^{3/2} g \, e^{-\epsilon_0/kT}.$$

Solving for μ/kT

$$\frac{\mu}{kT} = \ln \left[\left(\frac{h^2}{2\pi mkT} \right)^{3/2} \frac{n}{g} \right] + \frac{\epsilon_0}{kT}. \tag{129}$$

For the Maxwell–Boltzmann limit to hold we require

$$\ln \left[\left(\frac{h^2}{2\pi mkT} \right)^{3/2} \frac{n}{g} \right] \ll -1$$

which is the mathematical way of saying that phase space is not important. For electrons without internal excitations $g = 2$, and the previous expression can be re-expressed as

$$n_e \ll 2 \left(\frac{2\pi m_e kT}{h^2} \right)^{3/2} = 4.8 \times 10^{15} T^{3/2} \text{ cm}^{-3}$$

or

$$\rho \approx m_p n_e \ll 10^{-8} T^{3/2} \text{ g cm}^{-3}.$$

We can apply this criterion to a number of astrophysical environments to see whether the Maxwell–Boltzmann limit is justified. For the Sun, $T \approx 10^7$ K $\rightarrow$ $\rho \ll 300$ g cm^{-3} for complete non-degeneracy. The maximum density of the Sun is ≈ 100 g cm^{-3} so the criterion is largely satisfied. This is the case for most normal stars. The criterion is not satisfied in the case of white dwarfs which have degenerate cores. For the Universe as a whole when the electrons were non-relativistic ($T < 10^9$ K)

$$n_e \approx 10^{-5} \left(\frac{T}{2.7} \right)^3$$

$$\Rightarrow \frac{n_e}{T^{3/2}} = 10^{-5} \frac{T^{3/2}}{(2.7)^3} = \left(\frac{T}{10^9 \text{ K}} \right)^{3/2} 2 \times 10^7 \text{ cm}^{-3} \ll 10^{15}.$$

The more massive particles have even lower chemical potentials. For nucleons

$$n \ll 4.8 \times 10^{15} T^{3/2} \left(\frac{m_p}{m_e} \right)^{3/2} \approx 3.7 \times 10^{20} T^{3/2} \text{ cm}^{-3}$$

$$\rho \ll 6.1 \times 10^{-4} T^{3/2} \text{ g cm}^{-3} = 6.1 \times 10^8 \left(\frac{T}{10^8 \text{ K}} \right)^{3/2}. \tag{130}$$

This criterion is satisfied everywhere except in neutron stars. Thus neutron stars and white dwarfs are fully degenerate while some stars and Jovian planets are partially degenerate. As far as the Universe is concerned it can be considered non-degenerate which makes life a lot easier for describing particle reactions in the early Universe. Let us do that now.

5.5.2 *Chemical equilibrium*

Consider particle species, A, B, C, D. The reaction,

$$A + B \cdots \rightleftharpoons C + D \cdots$$

has the chemical potential relating as:

$$\mu_A + \mu_B \cdots \rightleftharpoons \mu_C + \mu_D \cdots$$

where μ_i is the chemical potential of species i and represents the energy change from adding one particle of type i.

5.5.3 The early Universe

Let us now consider the chemical equilibrium at $T = 10^{10}$ K. From the previous table we know that electrons and positrons are in equilibrium. Since the two particles interact via the pair production process we know that the following reaction will occur under equilibrium conditions

$$\gamma + \gamma \rightleftharpoons e^+ + e^-.$$

For photons which can be emitted and absorbed in unrestricted quantities, the chemical potential $\mu_\gamma = 0$. Therefore

$$\mu_{e^+} = -\mu_{e^-}.$$

We know from observations of the modern day Universe that

$$n_p \ll n_\gamma \approx n_{e^+} + n_{e^-}.$$

This means that the number density of electrons and positrons is very low relative to their number before they produced all the photons. Thus, their chemical potential, according to (129), must be relatively low. Thus, to a good approximation, $\mu_{e^+} \approx \mu_{e^-} \approx 0$. A similar statement can be made for the protons, neutrons and neutrinos.

5.5.4 The neutron–proton ratio

The current ratio of neutrons to protons in the Universe can be estimated by examining the nuclei of the elements. To a good approximation this ratio is given by comparing the abundances of H and He. Thus

$$\left[\frac{n}{p}\right] \approx \frac{1}{2}\left[\frac{He}{H}\right] \approx 12\%.$$

When the temperature of the Universe was 10^{10} K the neutrons and protons were in equilibrium with the electrons and neutrinos according to these reactions

$$n + e^+ \rightleftharpoons p + \bar{\nu}_e$$
$$n + \nu_e \rightleftharpoons p + e^-$$
$$n \rightarrow p + e^- + \bar{\nu}_e \quad \text{(one way)}.$$

The above is characterized by the chemical balance

$$\mu_n + \mu_{\nu_e} = \mu_p + \mu_{e^-}$$
$$\Rightarrow \mu_n = \mu_p. \tag{131}$$

Given the previous discussion these chemical potentials are also close to 0. However, let us use (131) to establish an important relationship between protons and neutrons. Combining (129) and (131) we have

$$\ln\left[\left(\frac{h^2}{2\pi m_n kT}\right)^{3/2}\frac{n_n}{g_n}\right] + \frac{\epsilon_{0,n}}{kT} = \ln\left[\left(\frac{h^2}{2\pi m_p kT}\right)^{3/2}\frac{n_p}{g_p}\right] + \frac{\epsilon_{0,p}}{kT}.$$

Now, $g_n \approx g_p \approx 2$. Also, $m_n \approx m_p$ is a good approximation for the slowly varying logarithmic term. Thus, this simplifies to

$$\ln n_n + \frac{\epsilon_{0,n}}{kT} = \ln n_p + \frac{\epsilon_{0,p}}{kT}$$
$$\Rightarrow \ln\left(\frac{n_n}{n_p}\right) = \frac{\epsilon_{0,p} - \epsilon_{0,n}}{kT} = -\frac{(m_n - m_p)c^2}{kT}$$

so that the ratio of neutron to proton number densities is given by

$$\frac{n_n}{n_p} = e^{-\Delta mc^2/kT}.$$

Now, $\Delta mc^2 = 0.78$ MeV $\rightarrow \Delta m/k = 9.1 \times 10^9$ K so that

$$\frac{n_n}{n_p} = e^{-9.1 \times 10^9/T}. \tag{132}$$

In order for this ratio to have the currently observed value of 0.12, the temperature in (132) must be

$$T = 4.3 \times 10^9 \text{ K}.$$

So at $T = 4.3 \times 10^9$ K the ratio froze out close to its current value. Why did that happen?

5.5.5 Reaction freeze-out

If the timescale of a chemical reaction is longer than the expansion timescale (over which the temperature cools sufficiently to change the reaction rate) then the reaction stops. We know that in the early Universe

$$\Omega = \frac{8\pi\rho G}{3H^2} \approx 1.$$

Thus

$$H = \frac{V}{R} = \frac{1}{R}\frac{dR}{dt} = \left(\frac{8\pi\rho G}{3}\right)^{1/2}.$$

We can now define the expansion timescale as

$$t_{exp} = \left|\frac{\rho_m}{d\rho_m/dt}\right|.$$

Since $\rho_m \propto R^{-3}$

$$t_{exp} = \frac{1}{3}\frac{R}{dR/dt} = \frac{1}{3H}$$

$$t_{exp} = \left(\frac{1}{24\pi\rho G}\right)^{1/2} = \frac{446 \text{ seconds}}{\sqrt{\rho_m/\text{g cm}^{-3}}}. \tag{133}$$

Equation (133) describes the non-relativistic expansion time. It is the characteristic dynamic timescale for a gravitating system.

Relativistic expansion

The first part of (133) is the same in the relativistic case so long as the ρ refers to the total mass-energy of the particles and radiation. For a single particle

$$E_i = \left(p_i^2 c^2 + m_i^2 c^4\right)^{1/2} \approx pc$$

for a highly relativistic particle. The equivalent ρ is therefore given by

$$\rho = \frac{1}{c^2}\sum_{vol} E_i = \frac{1}{c}\sum p_i.$$

We now define the Fermi energy density (i.e. the "mass") as

$$\rho_f = \sum 2\mathcal{N}p/c.$$

For the case: $\mu = 0$ we have

$$\rho_f = \frac{2}{c}\int \frac{d^3 p}{h^3}\frac{p}{[e^{pc/kT} + 1]}$$

$$\Rightarrow \rho_f c^2 = u_f = \frac{2c}{h^3}\int \frac{p}{[e^{pc/kT} + 1]}4\pi p^2\,dp$$

$$= \frac{8\pi}{h^3}\left(\frac{kT}{c}\right)^4\int_0^\infty \frac{x^3\,dx}{e^x + 1}. \tag{134}$$

For photons

$$u_\gamma = \frac{2}{ch^3} \int \frac{\mathrm{d}^3 p}{h^3} \frac{p}{e^{pc/kT} - 1}$$

$$= \frac{8\pi}{h^3} \left(\frac{kT}{c}\right)^4 \underbrace{\int_0^\infty \frac{x^3 \, \mathrm{d}x}{e^x - 1}}_{\pi^4/15}$$

$$\Rightarrow u_\gamma = \frac{8\pi^5 k^4}{15c^2 h^3} T^4 \equiv aT^4 \tag{135}$$

where a is the *Stefan–Boltzmann constant* $= 7.6 \times 10^{-15} \, \mathrm{erg \, cm^{-3} \, K^{-4}}$. Combining (134) and (135)

$$\frac{u_f}{u_\gamma} = \left[\frac{\int_0^\infty (x^3 \, \mathrm{d}x)/(e^x + 1)}{\int_0^\infty (x^3 \, \mathrm{d}x)/(e^x - 1)}\right] = \frac{7}{8}.$$

At $T > 10 \, \mathrm{MeV} \approx 10^{11} \, \mathrm{K}$

$$\rho_{tot} = \rho_\gamma + \rho_{e^+} + \rho_{e^-} + \rho_{\nu_e} + \rho_{\bar{\nu}_e} + \rho_{\nu_\mu} + \rho_{\bar{\nu}_\mu} + \rho_{\nu_\tau} + \rho_{\bar{\nu}_\tau}$$

$$= \frac{aT^4}{c^2}\left[1 + \frac{7}{8}(8)\right] = 8\frac{aT^4}{c^2}.$$

Thus, turning back to (133) with $\rho \to \rho_{tot}$ we have

$$t_{exp} = \frac{0.54 \, \mathrm{s}}{(T/10^6 \, \mathrm{K})^2} \approx \frac{0.5}{T_{10}^2} \, \text{seconds.} \tag{136}$$

5.5.6   Reaction timescale

The reaction timescale is defined by the expression

$$n\sigma v t \approx 1 \tag{137}$$

For the weak-force reaction equilibria

$$n + e^+ \to p + \bar{\nu}$$
$$n + \nu \to p + e^-$$

the reaction timescale is given by

$$\to t_{react} = \frac{1}{n_{e^+}\sigma_w c} \tag{138}$$

where

$$n_{e^+} \approx \frac{aT^4}{kT} \approx 5.5 \times 10^{31} T_{10}^3 \tag{139}$$

and

$$\sigma_w \approx 10^{-41} T_{10}^2 \text{ cm}^2. \tag{140}$$

Combining (138), (139) and (140)

$$t_{react} \approx \frac{0.06}{T_{10}^5} \text{ s}.$$

Comparing with (136) the condition $t_{react} > t_{exp}$ is satisfied when $0.06/T_{10}^5 > 0.5/T_{10}^2$. The timescales cross over at $T_{10} = 0.5$ and $t_{exp} = 2$ s.

Substituting this temperature into (132)

$$\frac{n_n}{n_p} = e^{-9.1 \times 10^9 / 5 \times 10^9} \approx 0.17$$

$$\rightarrow \frac{n_n}{n_n + n_p} \approx 0.14,$$

close to the abundances currently observed. However, free neutrons decay via

$$n \rightarrow p + e^+ + \bar{\nu}$$

on a timescale of 700 seconds. So why are there any neutrons left? Because stable nuclei formed and the neutrons were no longer free.

5.5.7 *Formation of deuterium*

Deuterium formed according to

$$n + p \rightleftharpoons D + \gamma.$$

At high temperatures the neutrons are fast and equilibrium is approached. Later T decreases and there is a "freeze-out". Early on the equilibrium is represented by

$$\mu_n + \mu_p = \mu_D + \mu_\gamma = \mu_D.$$

Substituting (129) into the above yields

$$\ln \left[\left(\frac{h^2}{2\pi m_n kT} \right)^{3/2} \frac{n_n}{g_n} \right] + \frac{\epsilon_{0n}}{kT} + \ln \left[\left(\frac{h^2}{2\pi m_p kT} \right)^{3/2} \frac{n_p}{g_p} \right] + \frac{\epsilon_{0p}}{kT}$$

$$= \ln \left[\left(\frac{h^2}{2\pi m_D kT} \right)^{3/2} \frac{n_D}{g_D} \right] + \frac{\epsilon_{0D}}{kT}.$$

Rearranging terms,

$$\left(\frac{n_n n_p}{n_D}\right) = \frac{g_n g_p}{g_D}\left[\frac{2\pi(m_n m_p/m_D)kT}{h^2}\right]^{3/2} e^{-\Delta E/kT} \tag{141}$$

where $\Delta E = E_{0n} + E_{0p} - E_{0D} = (m_n + m_p - m_D)c^2$. You may recognize the above equation as the *Saha* equation.

The Saha equation indicates that systems tend to break apart even when superficial estimates of binding energy suggest they should stay together. For the deuterium problem, $g_n = g_p = 2$, $g_D = 3$, $m_D = 2m_p = 2m_n$, $\Delta E = 8.07 + 7.29 - 13.14 = 2.22$ MeV, $\Delta E/k = 2.6 \times 10^{10}$ K.

Defining

$$R = \left(\frac{n_n n_p}{n_D}\right) \bigg/ \frac{1}{n_{nucleon}} \tag{142}$$

when $R \gg 1$ most nucleons are free, when $R \ll 1$ most nucleons are in D. For

$$n_N = n_{nucleon} = 10^{-5}\left(\frac{T}{2.7}\right)^3 = 5 \times 10^{20} T_9^3$$

combining (141) and (142), we get

$$R = \frac{3 \times 10^{48}}{T_9^{3/2}} e^{-26/T_9}.$$

From the above

$$R \approx 1 \qquad \text{when } T = 2.3 \times 10^8 \text{ K.}$$

Thus

$$t_{react} = \frac{1}{n\sigma v} \approx 3.4 \text{ seconds,}$$

while

$$t_{exp} = 1.2 \times 10^3 \text{ seconds.}$$

All the nucleons would be in D if it were not for other reactions that tend to destroy it, i.e.

$$D + p \rightarrow {}^3He\ldots$$

The early Universe is the site of exotic particle equilibria that lead up to the nucleosynthesis of the basic elements. The various particle equilibria and freeze-outs lead to the conditions we now see.

5.6 Further reading

Clayton, D. D. (1984) *Principles of Stellar Evolution and Nucleosynthesis*, University of Chicago Press, Chicago, USA.

Clayton, D. D. (2003) *From Hydrogen to Gallium*, Cambridge University Press, Cambridge, UK.

Combes, F., Mamon, G. A., Charmandaris, V. (Eds) (2000) *Dynamics of Galaxies: From the Early Universe to the Present*, Astronomical Society of the Pacific, San Francisco, USA.

Dekel, A. and Ostriker, J. (Eds) (1999) *Formation of Structure in the Universe*, Cambridge University Press, Cambridge, UK.

Fairall, A. P. (Ed) (1998) *Large-Scale Structures in the Universe*, John Wiley and Sons, New York, USA.

Harrison, E. (2000) *Cosmology: The Science of the Universe*, Cambridge University Press, Cambridge, UK.

Liddle, A. R. and Lyth, D. H. (2000) *Cosmological Inflation and Large Scale Structure*, Cambridge University Press, Cambridge, UK.

Mathews, G. J. (1988) *The Origin and Distribution of the Elements*, World Scientific, London, UK.

Narlikar, J. V. (1993) *Introduction to Cosmology*, Cambridge University Press, Cambridge, UK.

Peacock, J. A. (1999) *Cosmological Physics*, Cambridge University Press, Cambridge, UK.

Peebles, P. J. E. (1993) *Physical Cosmology*, Princeton University Press, Princeton, NJ, USA.

Rees, M. (2000) *New Perspectives in Astrophysical Cosmology*, Cambridge University Press, Cambridge, UK.

Songaila, A., Cowie, L. L., Hogan, C. J., Rugers, M. (1994) Deuterium Abundance and Background Radiation Temperature in High Redshift Primordial Clouds, *Nature*, **368**, 599.

Vangioni-Flam, E., Ferlet, R., Lemoine, M. (Eds) (2000) *Cosmic Evolution*, World Scientific, London, UK.

Chapter 6

Stellar structure and compact stars

Stars like the Sun owe their stability to a balance between gravitational and internal forces, the latter generated by thermonuclear reactions. This balance uniquely determines the internal structure of a star whose mass and chemical composition is known. The bulk of the gas inside such stars can be described through the ideal gas law. I will now present a very brief and very broad description of such stars. However, the real goal of this chapter is to describe very different kinds of stars, dead stars whose stability is owed not to thermonuclear reactions but to degenerate gas pressure. These stars are called white dwarfs and neutron stars. In doing so I will illustrate the interplay of gravitational dynamics and statistical mechanics.

6.1 Hydrostatic equilibrium

To be stable a star must be in mechanical equilibrium. It must not significantly expand or contract. Consider an idealized star of uniform density T and temperature ρ, radius R, mass M, as shown in Fig. 6.1. The gravitational pressure P_g can be obtained from the gravitational force F_g using the relation

$$P_g = \frac{F_g}{4\pi r^2}. \tag{143}$$

To calculate the net gravitational pressure we consider the interaction of thin shells of matter with mass interior to each shell and add up contributions from all shells. Thus

$$P_g = -\int_0^R G\frac{(4\pi r^2 \rho \, dr)\left(\frac{4}{3}\pi r^3 \rho\right)}{4\pi r^2 r^2} = -\frac{2}{3}\pi G\rho^2 R^2. \tag{144}$$

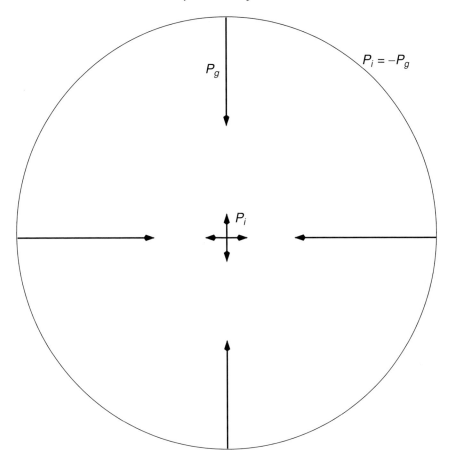

Fig. 6.1 Hydrostatic equilibrium. A star is stable when its internal gas pressure matches its gravitational pressure.

But

$$\rho = \frac{M}{\frac{4}{3}\pi R^3} \;\rightarrow\; \rho^2 = \frac{9}{16}\frac{M^2}{\pi^2 R^6}$$

$$\Rightarrow P_g = -\frac{3}{8\pi}G\frac{M^2}{R^4}. \tag{145}$$

The thermal pressure is provided by the kinetic motions of the gas particles and for a normal star follows the ideal gas law. Thus

$$P_i \approx nkT \qquad \text{(assume a pure hydrogen star)}$$

$$\approx \frac{MkT}{\frac{4}{3}\pi R^3 m_H}, \tag{146}$$

where m_H is the mass of the hydrogen atom.

For hydrostatic equilibrium to hold

$$P_i = -P_g.$$

Equating (145) and (146)

$$M = \frac{2RkT}{Gm_H}. \tag{147}$$

So, we see that an $M(R)$ relation is a consequence of hydrostatic equilibrium. Of course, real stars have gradients but this exercise serves to illustrate the basic concept. This particular $M(R)$ relation is dependent on the fact that the pressure is derived from thermonuclear reactions and arises from a quasi-ideal gas. Suppose the gas were degenerate? How would $M(R)$ change? This implication is discussed later in the chapter. What would happen if nuclear reactions were suddenly turned off? The star would begin to collapse and generate its energy by gravitational contraction. After some time we would expect the star to become sufficiently dense that gas degeneracy would begin to play a role. We can estimate at what radius that might happen as follows.

From the degeneracy condition (see below) for electrons

$$n \gg 5 \times 10^{15} T^{3/2} \rightarrow n > n_c = 5 \times 10^{16} T^{3/2}$$

where n is the number density, we can define a critical density, n_c, above which a gas is considered fully degenerate.

For $T = 10^7$ K, $n_c \approx 10^{27}$ cm^{-3}. But

$$n = \frac{M}{\frac{4}{3}\pi m_H R^3}$$

so that for $M = 1\, M_\odot$ and $m_H \approx 10^{-24}$ g

$$R < R_c \approx 10^9 \text{ cm} = 10^4 \text{ km}.$$

About the size of the Earth!

So what happens now? The answer requires us to work out a new hydrostatic equilibrium based on the degeneracy pressure.

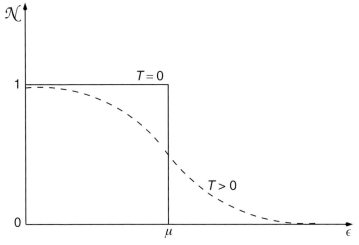

Fig. 6.2 The chemical potential. The chemical potential defines the energy at which the occupancies are falling off. At $T = 0$ the fall-off is abrupt. At $T > 0$ the fall-off is more gradual.

6.2 Fermion degeneracy

Fermi gases start to become degenerate when the finite volume of phase cells becomes important. Recall how the volume of a phase cell is defined

$$dV = \underbrace{dx\, dy\, dz}_{dV_s} \quad \underbrace{dp_x\, dp_y\, dp_z}_{dV_p} = h^3.$$

The total number of available phase cells is given by

$$N = \frac{2V_s V_p}{h^3} = \frac{8\pi}{3h^3} P_m^3 V_s \tag{148}$$

and is defined by the spatial volume, V_s, and the maximum momentum, P_m, in the ensemble. When the number of particles approaches the number of available phase cells the particle momenta are no longer determined by classical kinetics.

In a *differential* momentum volume

$$dN = \frac{2V}{h^3} 4\pi p_m^2\, dp = Z(p)\, dp \tag{149}$$

represents the distribution of states in p. Recall the Fermi–Dirac expression for the occupation number (105) which tells us what fraction of the available phase cells are filled at a given energy ϵ (Fig. 6.2). The distribution is characterized by the following limits.

Quantum limit: $\epsilon \ll \mu \rightarrow \mathcal{N}(\epsilon) = 1/(e^{-\mu/kT} + 1) \leq 1$

Classical limit: $\epsilon \gg \mu \rightarrow \mathcal{N}(\epsilon) = e^{-\epsilon/kT}$

For $T = 0$ $(\mu \gg kT) \rightarrow \mathcal{N}(\epsilon) = 1$, $(\epsilon < \mu)$ and $\mathcal{N}(\epsilon) = 0$, $(\epsilon > \mu)$.

From (149) the number density of available states is given by

$$\mathcal{Z}(p)\, \mathrm{d}p = \frac{Z(p)\, \mathrm{d}p}{V} = \frac{8\pi p^2}{h^3}\, \mathrm{d}p. \tag{150}$$

Comparing $\mathcal{Z}$ and $\mathcal{N}$, see (105) and (150), requires that common parameters be used. Let us proceed by converting $p \rightarrow \epsilon$ in (150).

Recall

$$\epsilon^2 = p^2 c^2 + m^2 c^4 \rightarrow \mathrm{d}\epsilon = c^2 \frac{p}{\epsilon}\, \mathrm{d}p.$$

Substituting into (150) yields

$$\mathcal{Z}(\epsilon)\, \mathrm{d}\epsilon = \frac{8\pi}{c^3 h^3}(\epsilon^2 - m^2 c^4)^{1/2}\epsilon\, \mathrm{d}\epsilon. \tag{151}$$

In the ultra-relativistic limit, $\epsilon \gg mc^2$

$$\mathcal{Z}(\epsilon)\, \mathrm{d}\epsilon = \frac{8\pi}{c^3 h^3}\epsilon^2\, \mathrm{d}\epsilon \qquad \text{(relativistic)}. \tag{152}$$

In the non-relativistic limit, $mc^2 \leq \epsilon \ll 2mc^2$

$$\Rightarrow (\epsilon^2 - m^2 c^4)^{1/2} = pc, \qquad \epsilon = \sqrt{p^2 c^2 + m^2 c^4} \approx mc^2 \left(1 + \frac{1}{2}\frac{p^2 c^2}{mc^2}\right)$$

$$\Rightarrow \mathcal{Z}(\epsilon)\, \mathrm{d}\epsilon \approx \frac{8\pi}{ch^3}mc^3 p\, \mathrm{d}\epsilon \approx \frac{4\pi}{h^3}(2m)^{3/2}\epsilon_k^{1/2}\, \mathrm{d}\epsilon \qquad \text{(non–relativistic)}. \tag{153}$$

Now we are in a position to determine the distribution of particle energies. We begin by defining the number density of particles as a function of energy

$$n(\epsilon)\, \mathrm{d}\epsilon = \mathcal{Z}(\epsilon)\, \mathrm{d}\epsilon\, \mathcal{N}(\epsilon). \tag{154}$$

At $T = 0$ we have

$$\rightarrow n(\epsilon)\, \mathrm{d}\epsilon = \begin{cases} \mathcal{Z}(\epsilon)\, \mathrm{d}\epsilon & \epsilon < \mu \\ 0 & \epsilon > \mu. \end{cases}$$

We can now determine various quantities like mean and average energy of an ensemble of particles at $T = 0$. For phase cells of energy, ϵ, the mean particle energy is defined as

$$\bar{\epsilon} = \epsilon \mathcal{N}(\epsilon) = \frac{\epsilon}{e^{(\epsilon - \mu)/kT} + 1} = \begin{cases} \epsilon, & \epsilon < \mu \\ 0, & \epsilon > \mu. \end{cases}$$

For all phase cells of all ϵ

$$\langle \epsilon \rangle = \frac{\int_0^\infty \mathcal{Z}(\epsilon) \mathcal{N}(\epsilon) \epsilon \, d\epsilon}{\int_0^\infty \mathcal{Z}(\epsilon) \mathcal{N}(\epsilon) \, d\epsilon} = \frac{\int_0^\infty n(\epsilon) \epsilon \, d\epsilon}{\int_0^\infty n(\epsilon) \, d\epsilon}. \tag{155}$$

For non-relativistic particles, from equations (153) and (155)

$$\langle \epsilon \rangle_{T=0} = \frac{\int_0^\mu \epsilon^{3/2}}{\int_0^\mu \epsilon^{1/2} \, d\epsilon} = \frac{3}{5}\mu = \frac{3}{5}\epsilon_F. \tag{156}$$

Here, we have defined $\mu = \epsilon_F$, the Fermi energy.

Similarly, for ultra-relativistic particles

$$\langle \epsilon \rangle_{T=0} = \frac{\int_0^\mu \epsilon^3 \, d\epsilon}{\int_0^\mu \epsilon^2 \, d\epsilon} = \frac{3}{4}\mu = \frac{3}{4}\epsilon_F. \tag{157}$$

We see that the chemical potential (the Fermi energy) is related to the mean energy of the particles. Let us now evaluate μ specifically.

The total number of particles in an ensemble is

$$n = \int_0^\infty n(\epsilon) \, d\epsilon = \int_0^\mu n(\epsilon) \, d\epsilon \qquad \text{at } T = 0$$

$$= \frac{8\pi}{c^3 h^3} \frac{\mu^3}{3} \qquad \text{(relativistic)}$$

$$= \frac{8\pi}{3h^3} (2m)^{3/2} \mu^{3/2} \qquad \text{(non-relativistic)}$$

so that

$$\mu = \left(\frac{3c^3 h^3}{8\pi} \right)^{1/3} n^{1/3} \qquad \text{(relativistic)} \tag{158}$$

$$\mu = \frac{1}{2m} \left(\frac{3h^3}{8\pi} \right)^{2/3} n^{2/3} \qquad \text{(non-relativistic)}. \tag{159}$$

In either case adding particles to the same volume increases the chemical potential (which is a way of defining the chemical potential).

6.2.1 White dwarf equation of state

These results make it possible to write down a relationship between pressure and density. From (156) and (157)

$$P = \frac{2}{3} \langle \epsilon \rangle_{T=0} \, n = \frac{2}{5} \mu n \qquad \text{(non-relativistic)}$$

$$P = \frac{1}{4} \mu n \qquad \text{(relativistic)}.$$

Using (158) and (159) to eliminate μ we get

$$P = \begin{cases} (1/8)(3c^3h^3/\pi)^{1/3} n^{4/3} & \text{(relativistic)} \\ (1/(10m))(3h^3/8\pi)^{2/3} n^{5/3} & \text{(non-relativistic)}. \end{cases} \qquad (160)$$

What do we notice about (160)? It does not depend on T! Remember this equation holds not only for $T = 0$ but for any T such that $\mu \gg kT$. Now let us make a white dwarf.

6.2.2 Mass–radius relation for white dwarfs

Since we want the star to be stable we begin with the hydrostatic equilibrium condition. From (145) and (160) we have

$$\frac{1}{10m_e} \left(\frac{3h^3}{8\pi} \right)^{2/3} n^{5/3} = \frac{3GM^2}{8\pi R^4} \qquad \text{(non-relativistic)}.$$

Using $n_H = M/(\frac{4}{3}\pi R^3 m_H) = n_e$

$$\Rightarrow M = \left[\frac{8\pi}{30Gm_e} \right]^3 \left[\frac{3h^3}{8\pi} \right]^2 \left[\frac{3}{4\pi m_H} \right]^5 R^{-3}$$

$$\Rightarrow M \approx 10^{60} R^{-3} \qquad \text{(non-relativistic)}. \qquad (161)$$

Repeating this for relativistic particles (a relativistic white dwarf)

$$M \approx \left(\frac{2\pi ch}{3G} \right)^{3/2} \left(\frac{3}{4\pi m_H} \right)^2 \approx 10^{34} \qquad \text{(relativistic)}. \qquad (162)$$

6.3 Internal structure of white dwarfs

We will now derive a more rigorous description of the structure of white dwarfs. First, we clarify the relationship between pressure and density, then the relationship

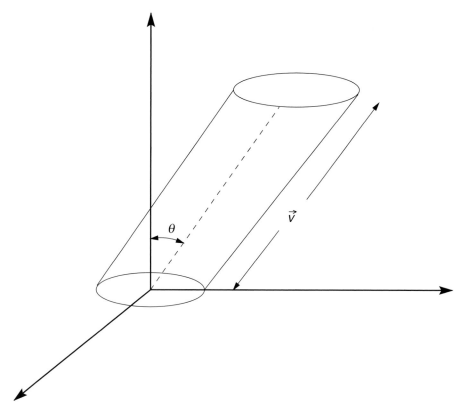

Fig. 6.3 Particle flux and the definitions of $\vec{v}$ and θ.

between pressure and mass density. This will set the stage for solving the equations of stellar structure.

6.3.1 Relationship between pressure and energy density

The microscopic approach

Consider Figs. 6.3 and 6.4. The momentum change of a single particle upon reflection off a surface is

$$\Delta p = -2p \cos \theta.$$

The pressure is the total change of momentum suffered by all particles incident on unit area in unit time

$$P = \int_0^\infty dv \int \Delta p v \cos \theta \, n(\Omega, v) \, d\Omega$$

$$= \int_0^\infty dv \int_0^{2\pi} d\phi \int_0^{\pi/2} \Delta p v \cos \theta \, n(\theta, \phi, v) \sin \theta \, d\theta$$

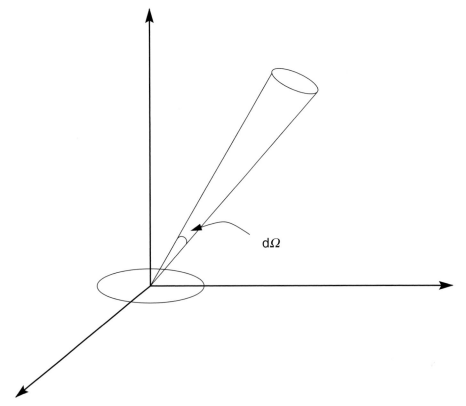

Fig. 6.4 Definition of the elemental solid angle.

where we have used the relation

$$[d\Omega = \sin\theta \, d\theta \, d\phi].$$

For an isotropic distribution

$$n(\theta, \phi, v) = \frac{n(v) \, dv}{4\pi}$$

so that

$$P = \int_0^\infty \frac{n(v)v \, dv}{4\pi} \int_0^{\pi/2} \Delta p \cos\theta \sin\theta \, d\theta \int_0^{2\pi} d\phi$$

$$= 2 \int_0^\infty \frac{n(v)v^2 \, dv}{4\pi} \int_0^{\pi/2} m \cos^2\theta \sin\theta \, d\theta \int_0^{2\pi} d\phi.$$

Now

$$2 \int_0^\infty \frac{n(v)v^2 \, dv}{4\pi} = \frac{1}{2\pi} n \langle v^2 \rangle$$

$$\int_0^{2\pi} d\phi = 2\pi$$

$$m \int_0^{\pi/2} \cos^2 \theta \sin \theta \, d\theta = -\frac{m}{3} \cos^3 \theta \Big|_0^{\pi/2} = \frac{m}{3}.$$

Combining

$$\Rightarrow P = \frac{1}{3} mn \langle v^2 \rangle = \frac{1}{3} \rho \langle v^2 \rangle = \frac{2}{3} U$$

$$\Rightarrow P = \frac{2}{3} U.$$

For a relativistic gas

$$m \langle v^2 \rangle \to \gamma m c^2$$

$$\Rightarrow P = \frac{1}{3} n \gamma m c^2 = \frac{1}{3} U$$

$$\Rightarrow P = \frac{1}{3} U.$$

Macroscopic approach

A star in hydrostatic equilibrium must satisfy

$$\frac{dP}{dr} = -\frac{\rho G M(r)}{r^2}.$$

Multiplying both sides by $V(r) \, dr = (4/3)\pi r^3 \, dr$

$$V(r) \, dP = -\frac{1}{3} 4\pi r^2 \rho \, dr \frac{G M(r)}{r} = -\frac{1}{3} \frac{G M}{r} \, dM$$

$$\Rightarrow \int_0^R V(r) \, dP = PV \Big|_0^R - \int_0^R P \, dV = -PV.$$

While

$$- \int \frac{1}{3} \frac{G M(r)}{r} \, dM = \frac{1}{3} \Omega.$$

We see that

$$PV = -\frac{1}{3} \Omega.$$

But, according to the Virial theorem

$$\Omega = -2 E_k$$

$$\Rightarrow P = \frac{2}{3} U \qquad (U = E_k / V),$$

where E_k is the kinetic energy.

For relativistic systems

$$E_k \approx \frac{1}{2} \sum m_i \langle v^2 \rangle n \rightarrow \frac{1}{2} \sum \gamma_i m_i n c^2$$

$$\Rightarrow P = \frac{1}{3} U \qquad \text{(relativistic)}.$$

Recall that for a degenerate gas

$$U = \frac{3}{5} \mu n \qquad \text{(non-relativistic)}$$

$$\Rightarrow P = \frac{2}{3} U = \frac{2}{5} \mu n$$

$$\Rightarrow P = \frac{1}{5 m_e} \left[\frac{3 h^3}{8 \pi} \right]^{2/3} n^{5/3} \qquad \text{(non-relativistic)}.$$

Or

$$U = \frac{3}{4} \mu n$$

$$\Rightarrow P = \frac{1}{3} U = \frac{1}{4} \mu n$$

$$P = \frac{1}{8} \left[\frac{3 c^3 h^3}{\pi} \right]^{1/3} n^{4/3} \qquad \text{(relativistic)}.$$

It is conventional to cast this equation into functions of mass and density.

6.3.2 *Relating electron number density to the mass density*

Atoms other than hydrogen must be taken into account when describing the internal structure of real white dwarfs. These atoms contribute multiple electrons and cannot therefore be neglected. Thus, the electron density is

$$n_e = \left(X + \frac{1}{2} Y + \frac{1}{2} Z \right) \frac{\rho}{m_H},$$

where X, Y, and Z are the fractional abundances of hydrogen, helium and heavy elements, respectively.
Since

$$X + Y + Z = 1$$

we have

$$n_e = \frac{1}{2} (1 + X) \frac{\rho}{m_H}$$

For $X = 0$

$$\Rightarrow n_e = \frac{\rho}{2 m_H}.$$

6.3.3 *Other sources of pressure*

Although usually minor, other sources of pressure include ions and photons

$$P_{\text{ion}} = \left(X + \frac{1}{4}Y \right) \frac{\rho k T}{m_H}$$

$$P_{\text{rad}} = \frac{U}{3} = \frac{n}{3} \langle h\nu \rangle = \frac{\sigma T^4}{3}.$$

6.3.4 *Equation of state*

Putting all of this together we have

$$P = \alpha \rho^{(n+1)/n}$$

where

$$\alpha = \begin{cases} (h^2/(20m_e m_H))(3/(\pi m_H))^{2/3}((1+X)/2)^{5/3} & \text{(non-relativistic)} \\ (hc/(8m_H))(3/(\pi m_H))^{1/3}((1+X)/2)^{4/3} & \text{(relativistic)} \end{cases}$$

$$n = \begin{cases} 3/2 & \text{(non-relativistic)} \\ 3 & \text{(relativistic)}. \end{cases}$$

Now we are set up to *properly* determine the internal structure of a white dwarf.

6.3.5 *Internal structure of white dwarfs*

There are three fundamental equations that govern the structure of white dwarfs.

Hydrostatic equilibrium In differential form we can describe the condition for hydrostatic equilibrium as

$$\frac{dP}{dr} = -\rho \frac{GM(r)}{r^2}. \tag{163}$$

Mass continuity

$$\frac{dM}{dr} = 4\pi r^2 \rho(r). \tag{164}$$

Equation of state

$$P = \alpha \rho^{(n+1)/n}. \tag{165}$$

Combining (163) and (164) yields

$$\frac{1}{r^2} \frac{d}{dr} \left(\frac{r^2}{\rho} \frac{dP}{d\rho} \right) = -4\pi G \rho(r). \tag{166}$$

It is mathematically expedient to write the density as

$$\rho(r) \equiv \rho_c \phi^n(r),$$ (167)

where $\rho_c = \rho(0)$ and $\phi(0) = 1$. Substituting (167) into (165) and (165) into (166) yields

$$(n+1)\alpha\rho_c^{1/n}\frac{1}{r^2}\frac{d}{dr}\left(r^2\frac{d\phi}{dr}\right) = -4\pi G\rho_c\phi^n.$$ (168)

We can further simplify things by defining the following parameters.

$$a \equiv \left[\frac{(n+1)\alpha\rho_c^{(1-n)/n}}{4\pi G}\right]^{1/2}$$ (169)

$$\zeta \equiv r/a.$$

Combining with (169) and (168) we now have

$$\frac{1}{\zeta^2}\frac{d}{d\zeta}\left(\zeta^2\frac{d\phi}{d\zeta}\right) = -\phi^n.$$ (170)

Equation (170) is known as the *Lane–Emden* equation.

Explicit solutions exist only for $n = 0, 1$ and 5. For our cases of interest ($n = 3/2$ and $n = 3$) there are no analytical solutions. Let us see what we can do.

Solving equation (170) requires boundary conditions for ϕ and $d\phi/d\zeta$. An obvious one for ϕ is

$$\phi(r = 0) = 1 \qquad (\rho(0) = \rho_c).$$

However, we also need one for $d\phi/d\zeta$. Let us see if we can determine what $(d\phi/d\zeta)|_{\zeta=0}$ should be.

In the vicinity of $\zeta = 0$ we can expand

$$\phi^n(\zeta) \approx \phi(0) + n\phi^{n-1}(0)(\zeta) = 1 + n\zeta\phi^{n-1}(0) = 1 + c\zeta$$ (171)

where c is a constant. Combining (171) with (164)

$$\Rightarrow dM \propto \zeta^2\rho_c(1+c\zeta) \propto \zeta^2 + c\zeta^3$$

$$\Rightarrow M \propto \zeta^3 + c'\zeta^4.$$ (172)

Substituting (172) into (163)

$$\Rightarrow \frac{dP}{dr} \propto \frac{\rho_c\phi^n(\zeta^3 + c'\zeta^4)}{\zeta^2} \rightarrow 0 \quad \text{as} \quad \zeta \rightarrow 0.$$

But $P = \alpha \rho^{(n+1)/n} = \alpha \rho_c \phi^{n+1}$ so that

$$\frac{dP}{dr} \propto (n+1)\phi^n \frac{d\phi}{dr} \Rightarrow \frac{d\phi}{d\zeta}(0) = 0.$$

Thus, the second boundary condition is

$$\left. \frac{d\phi}{d\zeta} \right|_{\zeta=0} = 0. \tag{173}$$

We can now get an approximate solution to (170) by expanding the solution in a power series for $\zeta < 1$

$$\phi(\zeta) \approx c_0 + c_1 \zeta + c_2 \zeta^2. \tag{174}$$

Substituting (174) into (170)

$$\frac{1}{\zeta^2} \frac{d}{d\zeta} \left(\zeta^2 \frac{d\phi}{d\zeta} \right) = -(c_0 + c_1\zeta + c_2\zeta^2)^n \approx -c_0^n \left[1 + n \left(\frac{c_1}{c_0}\zeta + \frac{c_2}{c_0}\zeta^2 \right) \right].$$

From (174), $d\phi/d\zeta \approx c_1 + 2c_2\zeta$ so that

$$\frac{1}{\zeta^2} \frac{d}{d\zeta}(c_1\zeta^2 + 2c_2\zeta^3) \approx -c_0^n \left[1 + n \left(\frac{c_1}{c_0}\zeta + \frac{c_2}{c_0}\zeta^2 \right) \right].$$

Simplifying

$$2c_1\zeta + 6c_2\zeta^2 \approx -c_0^n\zeta^2 - c_0^{n-1}nc_1\zeta^3 + \cdots$$

Equating the coefficients on the left- and right-hand sides, $c_0 = 1 \rightarrow c_1 = 0$, $c_2 = -1/6$, so that

$$\phi(\zeta) \approx 1 - \frac{1}{6}\zeta^2 \qquad \zeta < 1.$$

This solution satisfies both boundary conditions as can be readily verified. This procedure can be repeated with as many terms as needed for a given accuracy. However, the total solution, for *all* ζ requires the solution of (170) numerically for $\zeta > 1$. With the total solution it is possible to determine the mass and radius of the white dwarf given a central density.

6.3.6 *Estimating the radius and mass of a white dwarf*

The first zero of $\phi(\zeta)$, corresponding to $\phi(\zeta_0) = 0$, represents the radius of the white dwarf, see (169)

$$R = a\zeta_0. \tag{175}$$

The quantity

$$M(\zeta) = \int_0^{a\zeta} 4\pi r^2 \rho \, dr = \int_0^{a\zeta} 4\pi a^3 \rho_c \phi^n \zeta^2 \, d\zeta$$

can be combined with (170) to yield

$$M(\zeta) = -4\pi a^3 \rho_c \int_0^{\zeta} \frac{d}{d\zeta} \zeta^2 \frac{d\phi}{d\zeta} \, d\zeta = -4\pi a^3 \rho_c \zeta^2 \frac{d\phi}{d\zeta},$$

is the mass within a given ζ. The total mass of the star is therefore

$$M(\zeta_0) = -4\pi a^3 \rho_c \zeta_0^2 \left. \frac{d\phi}{d\zeta} \right|_{\zeta=\zeta_0}. \tag{176}$$

The *numerical solution* yields

n	ζ_0	$\zeta_0^2 (d\phi/d\zeta)_{\zeta=\zeta_0}$
3/2	3.65	−2.71
3	6.90	−2.02

For a relativistic white dwarf

$$R_{WD} = a\zeta_0 = 6.9a = 6.9 \left[\frac{\alpha \rho_c^{-2/3}}{\pi G} \right]^{1/2} \qquad \text{from (169)}$$

$$= 6.9 \left(\frac{\alpha}{\pi G} \right)^{1/2} \rho_c^{-1/3}.$$

From page 105

$$\alpha = \frac{hc}{8m_H} \left(\frac{3}{\pi m_H} \right)^{1/3} \left(\frac{1+X}{2} \right)^{4/3}.$$

For $X = 0$

$$\alpha = 4.8 \times 10^{14}$$

so that

$$R_{WD} = 3.3 \times 10^{11} \rho_c^{-1/3} \text{ cm.}$$

Now, we can estimate the mass from (176) and the previous table

$$M_{WD} = -4\pi a^3 \rho_c(-2.02) = 8\pi \left(\frac{\alpha}{\pi G} \right)^{3/2}.$$

$$= 2.76 \times 10^{33} \text{ g} = 1.4 \, M_\odot. \tag{177}$$

Equation (177) represents the *Chandrasekhar Limit.* Consider now a non-relativistic white dwarf. Referring to the previous table

$$R_{WD} = a\zeta_0 = 3.65a = 3.65 \left[\frac{5\alpha\rho_c^{-1/3}}{8\pi G} \right]^{1/2}$$

$$= 3.65 \left(\frac{5\alpha}{8\pi G} \right)^{1/2} \rho_c^{-1/6}.$$

From page 105

$$\alpha = \frac{h^2}{20m_e m_H} \left(\frac{3}{\pi m_H} \right)^{2/3} \left(\frac{1+X}{2} \right)^{5/3}.$$

For $X = 0$

$$\alpha = 3.1 \times 10^{12}$$
$$\Rightarrow R_{WD} = 1.1 \times 10^{10} \rho_c^{-1/6}$$
$$\Rightarrow R_{WD} = 0.16 R_\odot \rho_c^{-1/6}. \tag{178}$$

Repeating this for the mass

$$M_{WD} = 2.71 \, 4\pi a^3 \rho_c = 9 \times 10^{29} \rho_c^{1/2} \text{ g}$$
$$\Rightarrow M_{WD} = 4 \times 10^{-4} \rho_c^{1/2} M_\odot. \tag{179}$$

Combining (178) and (179)

$$\frac{M_{WD}}{M_\odot} = 1.6 \times 10^{-6} \left(\frac{R}{R_\odot} \right)^{-3} \qquad \text{(non-relativistic).} \tag{180}$$

The above *Mass–Radius* relation is valid for the mass range, $0.2 \leq M \leq 0.5 \, M_\odot$.

6.4 Stability of compact stars

We can investigate the stability of compact stars by analyzing the total energy of the star. In the process we will gain insight into why white dwarfs have an upper mass limit and we will set the stage for understanding neutron stars.

6.4.1 Total energy

The total energy of a star is given by the energy conservation law

$$E_T = E_G + E_k = -\frac{GM^2}{R_c} + UV.$$

Let us examine how the total energy scales with radius. We know that

$$V \propto R_c^3 \qquad \rho \propto R_c^{-3} \qquad U \propto \rho^\Gamma \propto R_c^{-3\Gamma}.$$

From the previous discussion we know that

$$\Gamma = 5/3 \qquad \text{non-relativistic}$$
$$\Gamma = 4/3 \qquad \text{relativistic}.$$

Thus, the scaling of the total energy is described by

$$E_T = -\frac{E_{g0}}{R_c} + \frac{E_{k0}}{R_c^{3(\Gamma-1)}}.$$

Thus, for non-relativistic gas

$$E_T = -\frac{E_{G0}}{R_c} + \frac{E_{k0}}{R_c^2}$$

for which there is a minimum at a finite radius. The minimum represents stability (Fig. 6.5).

In the relativistic case

$$E_T = -\frac{E_{G0}}{R_c} + \frac{E_{k0}}{R_c}$$

for which there is no minimum (Fig. 6.6).

Relativistic stars are unstable because the pressure does not increase fast enough with density to achieve hydrostatic equilibrium. Physically, on the microscopic scale, the relativistic electrons combine with the protons to form neutrons thereby reducing the available electrons for pressure support. The combination of the "soft" equation of state and the neutron production leads to the instability of relativistic stars. Let us look at the neutron production in a little more detail now because it is neutron production that leads to the formation of neutron stars.

6.4.2 Electron capture

In the laboratory neutrons are unstable and follow the usual decay process (Fig. 6.7)

$$n \rightarrow p + e^- + v + 0.78 \text{ MeV}.$$

In an environment in which the electrons are degenerate this process can only occur if there are available states for the created electron to occupy. If such states are not available, the decay does not take place. The highest energy electrons in a degenerate gas have

$$\epsilon_e = m_e c^2 + \epsilon_F.$$

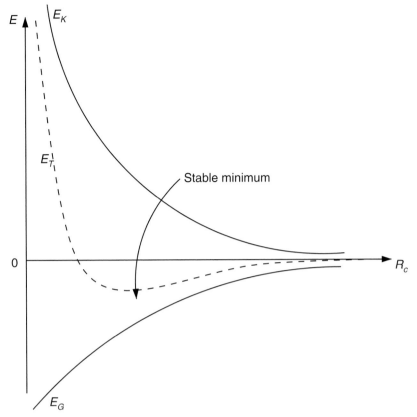

Fig. 6.5 Energy curve showing a stable point. When the total energy curve has a local minimum with respect to R the star has a stable point where it can exist in hydrostatic equilibrium.

When $\epsilon_F > 0.78 \, \mathrm{MeV} > m_e c^2 \approx 0.51 \, \mathrm{MeV}$, the electrons are just relativistic. Thus, the onset of electron capture is coincident with the equation of state becoming "soft" (Fig. 6.8).

6.4.3 Maximum density

Electrons are relativistic when

$$p_e \approx m_e c = \hbar k = \frac{h}{\lambda_e}$$

$$\Rightarrow \lambda_{ce} = \frac{h}{m_e c} = 2.4 \times 10^{-10} \, \mathrm{cm}$$

is the *Compton Wavelength*.

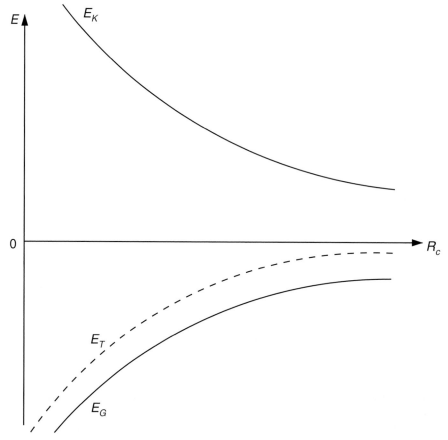

Fig. 6.6 Energy curve with no stable points. When the energy curve has no local minimum the star cannot attain hydrostatic equilibrium and it will collapse.

The mass density corresponding to a Compton wavelength and for matter with $A = 2Z$ is given by

$$\rho_{rel} \approx 2m_P \lambda_{ce}^{-3} \approx 2.3 \times 10^5 \text{ g cm}^{-3}.$$

A star of uniform density would have an upper density limit given by this density. Real stars, however, have density gradients so that the centers become relativistic before the other regions. Thus, for white dwarfs

$$(\rho_c)_{max} \approx 10^8 \text{ g cm}^{-3}.$$

As ρ increases past the above limit the ratio n/p grows, neutron matter forms and the pressure is dominated by neutrons, not electrons. The degenerate neutrons are non-relativistic, owing to their greater mass and the subsequent star is stable.

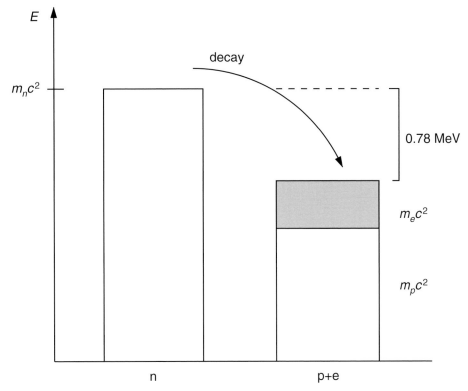

Fig. 6.7 Beta decay. Beta decay occurs when the rest mass energy of the neutron is greater than the combined energies of the electron and proton.

Thus

$$p_n \leq m_n c = \frac{h}{\lambda_{cn}}$$

$$\Rightarrow \lambda_{cn} \approx \frac{h}{m_{cn} c} \approx 1.3 \times 10^{-13} \text{ cm} = 1.3 \text{ fm}$$

$$\rho_{n,rel} \approx m_n / \lambda_{cn}^3 \approx 7 \times 10^{14} \text{ g cm}^{-3}.$$

The corresponding maximum density is then

$$\rho_{max} \approx \text{few} \times 10^{15} \text{ g cm}^{-3}.$$

At these densities nuclear reactions are strong and they play an important role in determining the structure of the neutron star. Consequently, the $P(\rho)$ relation for neutron stars is not well modeled and there are great debates over its exact form. Figure 6.9 summarizes the stability of compact stars.

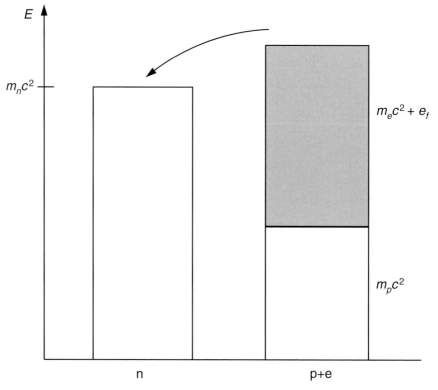

Fig. 6.8 Inverse beta decay (electron capture). Inverse beta decay occurs when the combined energies of the electron and proton are greater than the rest mass energy of the neutron. In this case it is the Fermi energy that causes the inverse process.

6.5 Structure of neutron stars

6.5.1 Overview

At the high densities and pressures characteristic of neutrons stars, the internal structure is quite interesting. At and just above the surface, where $P = 0$, ordinary matter can exist and there is no degeneracy. Just below the surface, it is believed that there is a relatively thin liquid layer, an ocean of non-relativistic degenerate matter. The interior is solid but the core may also be liquid as a result of a phase transition. The unusual structure is the result of the high internal densities which in turn is the result of the extremely strong gravitational potential.

The characteristic radius of a neutron star can be obtained from the maximum density just derived

$$\frac{4}{3}\pi \rho_{max} R_c^3 \approx 1 \, M_\odot$$

$$\Rightarrow R_c \approx 7 \times 10^5 \, \text{cm}.$$

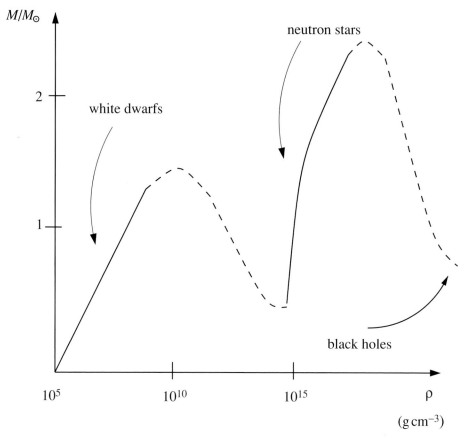

Fig. 6.9 Stability regimes for white dwarfs and neutron stars. The solid curves
indicate regions of stability. The dashed curves represent regions of instability for
white dwarfs and neutron stars.

For real stars, with density gradients

$$8 < R_{NS} < 20 \text{ km}.$$

The surface gravity is then given by

$$g_N = \frac{GM}{R^2} \approx 10^{14} \text{ cm s}^{-2}.$$

Compare that to the Earth where $g \approx 10^3$ cm s^{-2}.

6.5.2 Liquid layer

The liquid surface is the result of relatively low densities in the outer regions of
the neutron star. Here, there are still plenty of non-relativistic electrons and protons
and few neutrons. The situation is favorable (more later) for a liquid state when

$\rho < 10^6$ g cm^{-3}. We can get an approximate depth scale for the neutron star ocean using the following equilibrium equation

$$\frac{dP}{dz} = g\rho \tag{181}$$

which is just the equation of hydrostatic equilibrium where z is the depth. The equation of state for neutron stars is given by

$$P \approx 10^{13}\rho^{5/3} = k\rho^{5/3} \tag{182}$$

as one would expect for a degenerate non-relativistic star. Substituting (145) into (144) yields

$$dz = \frac{5}{3}\rho^{-1/3}\frac{k}{g}\,d\rho. \tag{183}$$

With the boundary condition $\rho(z = 0) = 0$ we get

$$z = \frac{5}{2}\frac{k}{g}\rho^{2/3} \qquad \text{for } (\rho < 10^6 \text{ g cm}^{-3})$$

$$\Rightarrow z = 25 \text{ meters} \left(\frac{\rho}{10^6 \text{ g cm}^{-3}}\right)^{2/3}. \tag{184}$$

We see that the liquid surface layer has a characteristic depth of ≈ 25 meters.

Why is the surface layer liquid?

In the outer regions of the neutron star the nuclei are surrounded by a sea of electrons (Fig. 6.10). The electrons are bound to the nuclei with an energy

$$E_c \approx -\frac{Ze^2}{r_i} \approx -\left(\frac{4\pi}{3}\right)^{1/3} Ze^2 n^{1/3}. \tag{185}$$

At high densities the binding energy is high and the particles form a crystal lattice. At lower densities the thermal energies of the particles can partially break down the lattice and form a liquid. The parameter that controls when that happens is given by

$$\Gamma = \frac{E_c}{kT}. \tag{186}$$

Computer simulations show that when $\Gamma < \Gamma_m \approx 150$, matter is liquid. The melting temperature T_m can then be defined by combining (185) and (186)

$$T_m = \frac{E_c}{k\Gamma_m} \approx 10^3\rho^{1/3}Z^{5/3} \text{ K} \qquad \text{for } A = 2Z$$

$$\Rightarrow T_m = 3 \times 10^7 \left(\frac{\rho}{10^6 \text{ g cm}^{-3}}\right)^{1/3} \tag{187}$$

for $Z \approx 30$.

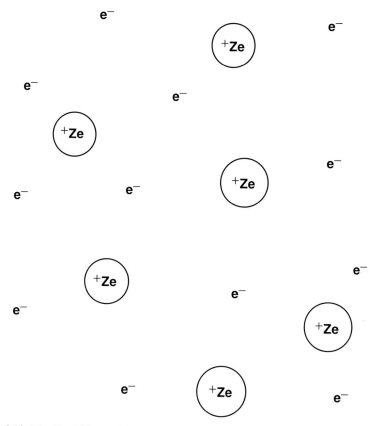

Fig. 6.10 The liquid layer. The ions are shown as circles. They are imbedded in a sea of electrons.

The internal temperature of a neutron star evolves with time so that

$$T_m > 10^8 \text{ K} \qquad t < 10^3 \text{ years}$$
$$T_m < 10^7 \text{ K} \qquad t > 10^6 \text{ years}.$$

According to (187) only a thin layer can be liquid after about a million years. That liquid has an estimated thickness of 10–100 meters. A solid crust forms below the ocean.

6.5.3 *The crust*

In the outer crust, electron capture alters the composition according to

$$e^- + (Z, N) \to (Z - 1, N + 1) + \nu.$$

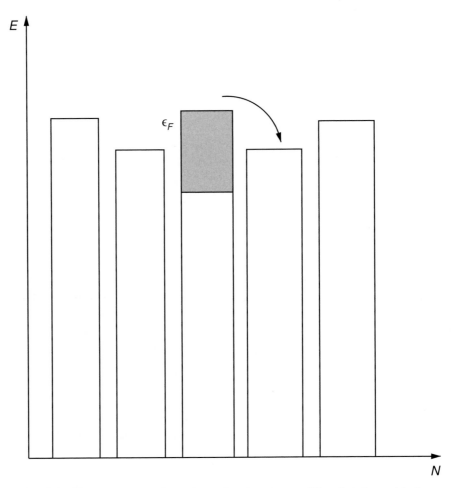

Fig. 6.11 Electron capture brought on by degeneracy. The plot shows binding energies for nuclei of a fixed atomic mass. The presence of a biasing Fermi energy allows electron capture to occur to the right of the minimum.

Consider the plot of binding energies for nuclei of a fixed atomic mass in Fig. 6.11. Electron capture should not be possible to the right of the minimum. However, the high Fermi energies allow this to happen.

As the density increases with depth there is a corresponding increase in the Fermi energy, ϵ_F. At high enough Fermi energies of $\epsilon_F \approx 20$ MeV and $\rho \approx 10^{12}$ g cm^{-3} neutrons begin to *drip* out of the nuclei according to

$$(Z, N) \rightarrow (Z, N - 1) + \mathrm{n}.$$

Thus, the outer crust has an altered composition of electrons and nuclei while the inner crust is a mixture of neutrons, nuclei and electrons.

6.5.4 The core

In the core the density approaches $\rho \approx 2.8 \times 10^{14}$ g cm^{-3}, at which point the nuclei are merging together to form a liquid core. This liquid consists of free neutrons, protons and electrons. At this point the Fermi energies are approaching 100 MeV, which is the rest mass energy of a muon. At the very center muons become a part of the mixture. For $T < T_c \approx 10^{9-10}$K, free neutrons in the crust and in the core form a neutron superfluid. Free protons in the core form a superconductor. The superfluid represents flow without viscosity while superconductivity represents energy conduction without losses.

6.6 Pulsars

Pulsars, as the name suggests, emit repeated pulses of radiation. The frequency of pulsation is rapid ranging from milliseconds to seconds of time. The pulse frequency is also very regular. The combination of rapid pulsation and regularity is a severe constraint on the physics of these objects. There are two possibilities. Pulsars are either dynamical pulsators (they contract and expand periodically like pulsating variable stars) or rapidly rotating objects.

The dynamical timescale for a stellar oscillation is given by

$$t_{dyn} \approx (24\pi G\rho)^{-1/2}$$

which is related to the free fall time. The period of oscillation must exceed the dynamical timescale, otherwise the star is destroyed.

$$P > t_{dyn} = \frac{446}{\sqrt{\rho}} \text{ seconds.}$$

If we use a typical period for a pulsar (say that of the Crab nebula) we can derive a lower limit on the density of the star.

$$\rho > \left(\frac{446}{0.033 \text{ s}}\right)^2 \approx 2 \times 10^8 \text{ g cm}^{-3}$$

which is just above the maximum density of a white dwarf. For faster pulsars, white dwarfs can be ruled out.

In the case of rotation

$$a_{cent} = \Omega^2 R \qquad a_{grav} = \frac{GM}{R^2}$$

and for the star to be stable against disruption

$$\Omega^2 R = \left(\frac{2\pi}{P}\right)^2 R < \frac{GM}{R^2}$$

$$\Rightarrow (2\pi)^2 P^{-2} < \frac{GM}{R^3} = \frac{4}{3} G\pi\rho$$

$$P > \left(\frac{3\pi}{G\rho}\right)^{1/2} = \frac{1200}{\sqrt{\rho}}$$

$$\Rightarrow \rho > \left(\frac{1200}{P}\right)^2 = 1.3 \times 10^9 \text{ g cm}^{-3}.$$

The density has to be greater than $\approx 10^9$ g cm^{-3}, especially for the faster millisecond pulsars. This range of densities completely eliminates white dwarfs as candidates. Thus it seems that regardless of whether pulsars are oscillating or rotating they are probably neutron stars. So which is it? Are they oscillating or rotating?

The periods of pulsars can be measured so accurately that it is possible to measure very small changes in pulse periods. In the case of the Crab pulsar

$$P = 0.03342 \ldots \text{ s}$$

$$\dot{P} = 2.5 \times 10^{-12} > 0$$

$$\Rightarrow \frac{P}{\dot{P}} = 2650 \text{ years.}$$

The deceleration timescale is 2650 years. All pulsars for which the deceleration timescales have been measured have positive $\dot{P}$. That is, they all slow down and increase their periods.

If pulsars were oscillating sources their evolution would be determined by the Virial theorem. Since they are emitting radiation and since they have no internal heat generation their density should increase with time. But according to these equations

$$P_{osc} \propto \rho^{-1/2}$$

which predicts that the period should *decrease* with time, contrary to what is actually observed.

A rotating star loses rotational energy as it radiates so that

$$\frac{d}{dt}\left(\frac{I}{2}\Omega^2\right) = \Omega \dot{\Omega} I + \frac{\Omega^2}{2}\frac{dI}{dt} < 0$$

if $I \approx$ constant. Thus

$$\dot{\Omega} < 0 \Rightarrow \dot{P} > 0$$

as is observed.

Pulsars are rapidly rotating neutron stars. We will discuss pulsars in more detail later, when we have reviewed EM. The radiation pulsars give off contains some very interesting physics as we will see in Section 9.6.

6.7 Further reading

Clayton, D. D. (1984) *Principles of Stellar Evolution and Nucleosynthesis*, University of Chicago Press, Chicago, USA.

Frank, J., King, A., Raine, D. (2002) *Accretion Power in Astrophysics*, Cambridge University Press, Cambridge, UK.

Glendenning, N. K. (1996) *Compact Stars: Nuclear Physics, Particle Physics and General Relativity*, Springer-Verlag, Heidelberg, Germany.

Kaminker, A. M. (1995) *The Physics of Neutron Stars*, Nova Science Publishers, Hauppauge, NY, USA.

Riffert, H. (Ed) (1996) *Matter at High Densities in Astrophysics*, Springer-Verlag, Heidelberg, Germany.

Shapiro, S. L. and Teukolsky, S. A. (1983) *Black Holes, White Dwarfs and Neutron Stars*, John Wiley and Sons, New York, USA.

Part III

Electromagnetism

Observational astronomy is based on the detection of radiation emitted by astro-physical objects. All morphological and physical information about astrophysical sources is derived from observation and analysis of the emitted radiation. Continuum radiation is a natural consequence of the principle that accelerating charges radiate. In this part of the book I will review and apply principles of electromagnetism to further our understanding of important astrophysical phenomena, demonstrating in the process that important radiation processes can be derived from the basic principle that accelerating charges radiate. To that end, I will develop the theory that describes bremsstrahlung and synchrotron radiation. A theoretical understanding of these two radiation mechanisms allows us to interpret the emission of a wide range of objects, ranging from distant radio galaxies to nearby HII regions. The specific astrophysical topics covered in this part of the book include: the radiative properties of pulsars, dispersion and Faraday rotation of electromagnetic radiation, active binary star systems, accretion disks, supernova remnants, particle acceleration, cosmic rays and active galaxies.

Chapter 7

Radiation from accelerating charges

A proper interpretation of the radiation detected from astrophysical sources requires us to understand why radiation occurs and what the main radiation mechanisms are. In this chapter I derive the Lienard–Wiechert (L–W) potentials, which form the basis for describing radiating charges using basic principles of EM and relativity. The L–W potentials are used to describe two of the most important continuum radiation mechanisms in astronomy, thermal bremsstrahlung and synchrotron radiation.

7.1 The Lienard–Wiechert potential

7.1.1 Scalar and vector potentials

Recall the vector relation

$$\vec{\nabla} \cdot (\vec{\nabla} \times \vec{a}) = 0$$

from which we can define a vector potential such that

$$\vec{\nabla} \cdot \vec{B} = \vec{\nabla} \cdot (\vec{\nabla} \times \vec{A}) = 0$$
$$\Rightarrow \vec{B} = \vec{\nabla} \times \vec{A}, \tag{188}$$

where $\vec{B}$ is the magnetic field and $\vec{A}$ is the associated vector potential. Substituting (188) into Faraday's Law (from Maxwell's equations (242)–(245))

$$\vec{\nabla} \times \vec{E} = -\frac{1}{c}\frac{\partial}{\partial t}(\vec{\nabla} \times \vec{A}) = -\frac{1}{c}\vec{\nabla} \times \frac{\partial \vec{A}}{\partial t},$$

where $\vec{E}$ is the electric field and ϕ is the associated scalar potential. But recall that $\vec{\nabla} \times (\vec{\nabla}\phi) = 0$ which we are then free to add to the right-hand side

$$\Rightarrow \vec{\nabla} \times \vec{E} = -\vec{\nabla} \times \left[\frac{1}{c}\frac{\partial \vec{A}}{\partial t} + \vec{\nabla}\phi \right]$$

$$\Rightarrow \vec{E} = -\frac{1}{c}\frac{\partial \vec{A}}{\partial t} - \vec{\nabla}\phi. \tag{189}$$

Generalizing this result

$$\vec{A}' = \vec{A} + \vec{\nabla}\psi$$
$$\phi' = \phi - \frac{1}{c}\frac{\partial\psi}{\partial t},$$

These equations represent *gauge transformations*. The gauge transformations give us mathematical flexibility for solving various EM problems. They are useful because they do not change the underlying physics. A particularly useful gauge for dealing with radiation is given by the *Lorentz condition*

$$\vec{\nabla}\cdot\vec{A} - \frac{1}{c}\frac{\partial\phi}{\partial t} = 0. \tag{190}$$

Let us now substitute (188) and (189) into Coulomb's and Ampere's Laws (see Maxwell's equations (242)–(245))

$$\Rightarrow -\nabla^2\phi - \frac{1}{c}\frac{\partial}{\partial t}(\vec{\nabla}\cdot\vec{A}) = 4\pi\rho_e \tag{191a}$$

$$\vec{\nabla}\times(\vec{\nabla}\times\vec{A}) = \frac{4\pi}{c}\vec{j}_e - \frac{1}{c}\vec{\nabla}\frac{\partial\phi}{\partial t} - \frac{1}{c^2}\frac{\partial^2\vec{A}}{\partial t^2}. \tag{191b}$$

Combining (190) and (191) we obtain

$$\Box^2\begin{pmatrix}\phi\\\vec{A}\end{pmatrix} = -4\pi\begin{pmatrix}\rho_e\\\vec{j}_e/c\end{pmatrix}. \tag{192}$$

This is known as *the inhomogeneous wave equation* where

$$\Box^2 \equiv \nabla^2 - \frac{1}{c^2}\frac{\partial^2}{\partial t^2}.$$

It is possible to solve (192) with the help of Green's function.

7.1.2 Green's function solution

The first step is to solve the special case:

$$\Box^2 G(\vec{x}, t; \vec{x}', t') = 4\pi\delta(x - x')\delta(t - t') \tag{193}$$

for G, then

$$\phi(\vec{x}, t) = \int G(\vec{x}, t; \vec{x}', t')\rho_e(\vec{x}', t')\, dx'\, dt' \tag{194a}$$

$$\vec{A}(\vec{x}, t) = \int G(\vec{x}, t; \vec{x}', t')\frac{\vec{j}_e(\vec{x}', t')}{c}\, d^3x'\, dt'. \tag{194b}$$

The solution to (193) is given by

$$G_{ret} = \frac{\delta(t' - (t - |\vec{x} - \vec{x}'|/c))}{|\vec{x} - \vec{x}'|} \tag{195}$$

so that (194) becomes

$$\phi(\vec{x}, t) = \int \frac{\delta[t' - (t - |\vec{x} - \vec{x}'|/c)]}{|\vec{x} - \vec{x}'|} \rho_e(\vec{x}', t') \, d^3x' \, dt'$$

$$\vec{A}(\vec{x}, t) = \int \frac{\delta[t' - (|\vec{x} - \vec{x}'|/c)]}{|\vec{x} - \vec{x}'|} \frac{\vec{j_e}}{c} \, d^3x' \, dt'.$$

For $t' = t - |\vec{x} - \vec{x}'|/c$, the *retarded time*, we have

$$\phi(\vec{x}, t) = \int \frac{\rho_e(\vec{x}', t')}{|\vec{x} - \vec{x}'|} \, d^3x' \tag{196a}$$

$$\vec{A}(\vec{x}, t) = \frac{1}{c} \int \frac{\vec{j_e}(x', t')}{|\vec{x} - \vec{x}'|} \, d^3x'. \tag{196b}$$

These are known as the *retarded potentials*.

7.1.3 The L–W potentials

Consider a single charge q moving on a trajectory $\vec{r}(t)$

$$\Rightarrow \rho_e = q\delta[\vec{x} - \vec{r}(t)] \tag{197a}$$

$$\vec{j_e} = q\vec{v}\delta[\vec{x} - \vec{r}(t)]. \tag{197b}$$

Then, in terms of the preceding approach

$$\begin{bmatrix} \rho_e(\vec{x}', t') \\ \vec{j_e}(\vec{x}', t')/c \end{bmatrix} = \int \begin{bmatrix} q \\ q\vec{v}(\tau)/c \end{bmatrix} \delta(\vec{x}' - \vec{r}(\tau))\delta(\tau - t') \, d\tau.$$

Substituting into (196) and integrating over $\vec{x}'$

$$\phi(\vec{x}, t) = \int \int \frac{q\delta(\vec{x}' - \vec{r}(\tau))}{|\vec{x} - \vec{x}'|} \delta(\tau - t') \, d^3x' \, d\tau$$

which is nonzero only for $\vec{x}' = \vec{r}(\tau)$

$$\Rightarrow \phi(\vec{x}, t) = \int \frac{q\delta(\tau - t')}{|\vec{x} - \vec{r}(\tau)|} \, d\tau. \tag{198}$$

Similarly

$$\Rightarrow \vec{A}(\vec{x}, t) = \int \frac{q\vec{v}(\tau)/c}{|\vec{x} - \vec{r}(\tau)|} \delta(\tau - t') \, d\tau. \tag{199}$$

Now let us define $\vec{R}(\tau) \equiv \vec{x} - \vec{r}(\tau)$

$$\Rightarrow t' = t - |\vec{x} - \vec{x}'|/c = t - |\vec{x} - \vec{r}(\tau)|/c \tag{200}$$

$$t' = t - R(\tau)/c. \tag{201}$$

Now let $\tau' \equiv \tau - t' = \tau - t + R(\tau)/c$

$$\Rightarrow d\tau' = [1 + \dot{R}(\tau)/c] \, d\tau. \tag{202}$$

After some manipulation

$$\frac{d\tau}{R(\tau)} = \frac{d\tau'}{R(\tau) - \vec{R}(\tau) \cdot \vec{v}(\tau)/c}. \tag{203}$$

Substituting (203) into (198) and (199)

$$\begin{pmatrix} \phi(\vec{x}, t) \\ \vec{A}(\vec{x}, t) \end{pmatrix} = \int \begin{pmatrix} q \\ q\vec{v}(\tau)/c \end{pmatrix} \frac{\delta(\tau') \, d\tau'}{R(\tau) - \vec{R}(\tau) \cdot \vec{v}(\tau)/c}. \tag{204}$$

Equation (204) is nonzero only for $\tau' = 0$ so

$$\phi(\vec{x}, t) = \frac{q}{R(\tau) - \vec{R}(\tau) \cdot \vec{v}/c} \Big]_{ret} \tag{205a}$$

$$\vec{A}(\vec{x}, t) = \frac{q\vec{v}(\tau)/c}{R(\tau) - \vec{R}(\tau) \cdot \vec{v}/c} \Big]_{ret} \tag{205b}$$

where $]_{ret}$ indicates that quantities are evaluated at the retarded time, that is, for

$$\tau = t - R(\tau)/c. \tag{206}$$

Equations (205) and (206) represent the Lienard–Wiechert potentials.

7.2   Electric and magnetic fields of a moving charge

The $\vec{E}$ and $\vec{B}$ fields can be evaluated according to

$$\vec{E} = -\vec{\nabla}\phi - \frac{1}{c}\frac{\partial \vec{A}}{\partial t} \quad \text{and} \quad \vec{B} = \vec{\nabla} \times \vec{A}. \tag{207}$$

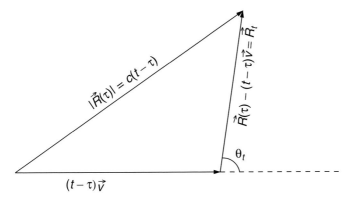

Fig. 7.1 Trajectory of a charge moving at constant velocity.

Now, from (206)

$$\frac{\partial \tau}{\partial t} = \frac{1}{1 + \dot{R}/c} = \frac{1}{1 - \hat{R} \cdot \vec{v}/c} = \frac{1}{\kappa}. \tag{208}$$

Combining (207), (208) and (205)

$$\vec{E} = \frac{q}{(R - \vec{R} \cdot \vec{v}/c)^3} \left[\left(1 - \frac{v^2}{c^2} \right) \left(\vec{R} - \frac{R\vec{v}}{c} \right) + \frac{\vec{R}}{c^2} \times \left[\left(\vec{R} - \frac{R\vec{v}}{c} \right) \times \dot{\vec{v}} \right] \right] \tag{209}$$

$$\vec{B} = \hat{R} \times \vec{E}.$$

Note that the first term in (209) $\propto 1/R^2$ and yields the *Coulomb field*. The second term $\propto 1/R$ and represents the *Radiation field*.

7.2.1 Moving charge at constant velocity

When $\dot{\vec{v}} = 0$ only Coulomb fields are possible. From Fig. 7.1

$$R - \frac{\vec{R} \cdot \vec{v}}{c} = R_t \left(1 - \frac{v^2}{c^2} \sin^2 \theta_t \right)^{1/2}$$

$$= \left[|\vec{R}_t|^2 - \left| \vec{R}_t \times \frac{\vec{v}}{c} \right|^2 \right]^{1/2}. \tag{210}$$

For a proof see Shu (1992). Combining (207), (209) and (210)

$$\Rightarrow \vec{E} = \frac{q(1 - (v^2/c^2))\vec{R}_t}{(|\vec{R}_t|^2 - |\vec{R}_t \times (\vec{v}/c)|^2)^{3/2}} \quad \text{and} \quad \vec{B} = \frac{\vec{v}}{c} \times \vec{E}. \tag{211}$$

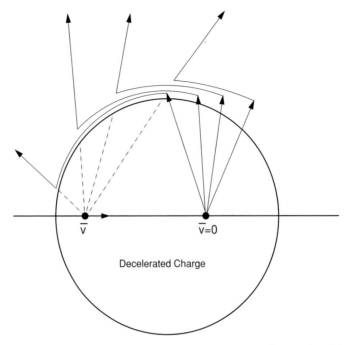

Fig. 7.2 The fields of a decelerating charge. A charge is moving initially at constant speed $\vec{v}$. It decelerates instantly to $\vec{v} = 0$. Portions of the resulting E fields are shown.

Special cases

$$\vec{v} = 0 \; \rightarrow \; \vec{E} = \frac{q}{R_t^2} \, \hat{r}$$

$$\vec{v} \perp \vec{R} \; \rightarrow \; \vec{E} = \frac{\gamma q}{R_t^2} \, \hat{r}$$

$$\vec{v} \parallel \vec{R} \; \rightarrow \; \vec{E} = \frac{1}{\gamma^2} \frac{q}{R_t^2} \, \hat{r}.$$

A qualitative picture based on (211) is shown in Fig. 7.2. Inspection of Fig. 7.2 shows that when a moving charge suddenly stops, corresponding to an instantaneous acceleration, the static fields form two separate regimes. Causality demands that the fields associated with the two regimes be connected, as shown. Given the finite speed of light, an external observer first sees the initial field associated with the moving charge. After a finite amount of time, corresponding to the time it takes for the field of the stationary charge to reach the observer, a discontinuity in the E and B fields sweeps over the observer. This discontinuous field is the EM wave associated with the acceleration of the charge. Note that the $\vec{E}$ field lines along the discontinuity are orthogonal to the direction of propagation and that the discontinuity propagates at the speed of light, both characteristics of propagating EM radiation. Causality demands that accelerating charges radiate.

7.2.2 Radiation from accelerating charges – the far zone

Let us now investigate the second term in (209). In the far field, $\vec{R} = \hat{k}x$.

$$\Rightarrow \vec{E} = \frac{q}{xc^2\kappa^3} \left\{ \hat{k} \times \left[\left(\hat{k} - \frac{\vec{v}}{c} \right) \times \dot{\vec{v}} \right] \right\} \tag{212}$$

$$\vec{B} = \hat{k} \times \vec{E}.$$

For $v/c \ll 1 \rightarrow |\vec{E}| = (q|\dot{\vec{v}}|\sin\theta)/xc^2$.

For $v/c \approx 1 \rightarrow \kappa = 1 - (\hat{k} \cdot \vec{v})/c \approx 1 - \cos\theta$, demonstrating that there is relativistic beaming around $\theta \approx 0$.

7.2.3 Angular distribution of radiation

Recall that

$$\frac{dP}{d\Omega} = x^2 \times \underbrace{|\vec{S}|}_{\text{Poynting Vector}} = \frac{c}{4\pi}(x|\vec{E}|)^2. \tag{213}$$

Define

$$\vec{g} = \frac{1}{\kappa^3} \left\{ \hat{k} \times \left[\left(\hat{k} - \frac{\vec{v}}{c} \right) \times \dot{\vec{v}} \right] \right\} \tag{214}$$

with the help of Fig. 7.3, such that, from (212), (213) and (214)

$$\frac{dP}{d\Omega} = \frac{q^2|\vec{g}|^2}{4\pi c^3}.$$

Expanding the double product in (212)

$$\vec{g} = \frac{1}{\kappa^3} \left[(\hat{k} \cdot \dot{\vec{v}}) \left(\hat{k} - \frac{\vec{v}}{c} \right) - \kappa \dot{\vec{v}} \right]$$

$$\Rightarrow \vec{g} \cdot \vec{g} = g^2 = \frac{1}{\kappa^4}|\dot{\vec{v}}|^2 + \frac{2}{\kappa^5}(\hat{k} \cdot \dot{\vec{v}}) \left(\dot{\vec{v}} \cdot \frac{\vec{v}}{c} \right) - \frac{1}{\kappa^6} \left(1 - \frac{v^2}{c^2} \right) (\hat{k} \cdot \dot{\vec{v}})^2 \tag{215}$$

where we have used $\kappa = 1 - (\hat{k} \cdot \vec{v})/c$. We also note that

$$\hat{k} \cdot \frac{\vec{v}}{c} = \frac{v}{c}\cos\theta$$

$$\hat{k} \cdot \dot{\vec{v}} = |\dot{\vec{v}}|(\sin\theta\cos\phi\sin i + \cos\theta\sin\phi\cos i)$$

$$\dot{\vec{v}} \cdot \frac{\vec{v}}{c} = |\dot{\vec{v}}|\frac{v}{c}\cos i. \tag{216}$$

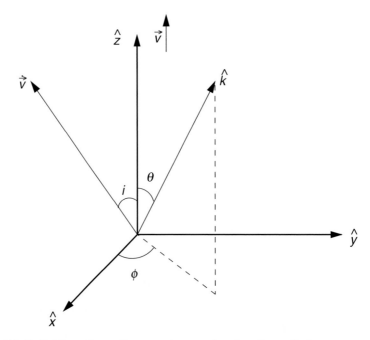

Fig. 7.3 Definition of coordinate system used to describe radiation patterns.

Armed with (215) and (216), two special cases can now be described.

Case 1 $(\vec{v} \| \dot{\vec{v}})$

$$\vec{v} \| \dot{\vec{v}}(i = 0) \Rightarrow \hat{k} \cdot \frac{\vec{v}}{c} = \frac{v}{c} \cos\theta \quad \text{and} \quad \hat{k} \cdot \dot{\vec{v}} = |\dot{\vec{v}}| \cos\theta \quad \text{and} \quad \dot{\vec{v}} \cdot \frac{\vec{v}}{c} = |\dot{\vec{v}}|\frac{v}{c}.$$

Substitute into (215)

$$\Rightarrow \frac{g^2}{|\dot{\vec{v}}|^2} = \frac{\sin^2\theta}{(1 - (v/c)\cos\theta)^6}.$$

The radiation pattern, corresponding to this equation, is shown in Fig. 7.4.

Case 2 $(\dot{\vec{\perp}}\dot{\vec{v}})$

$$\vec{v} \perp \dot{\vec{v}}(i = 90°) \Rightarrow \hat{k} \cdot \frac{\vec{v}}{c} = 0 \quad \text{and} \quad \hat{k} \cdot \dot{\vec{v}} = |\dot{\vec{v}}|(\sin\theta \cos\phi) \quad \text{and} \quad \dot{\vec{v}} \cdot \frac{\vec{v}}{c} = 0$$

$$\Rightarrow \frac{g^2}{|\dot{\vec{v}}|^2} = \left(1 - \frac{v}{c}\cos\theta\right)^{-4} - \left(1 - \frac{v^2}{c^2}\right)\left(1 - \frac{v}{c}\cos\theta\right)^{-6}\sin^2\theta\cos^2\phi.$$

The radiation pattern, corresponding to this equation, is shown in Fig. 7.5.

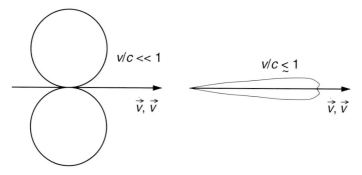

Fig. 7.4 Radiation pattern when $\dot{\vec{v}} \parallel \vec{v}$.

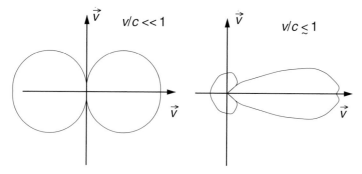

Fig. 7.5 Radiation pattern when $\dot{\vec{v}} \perp \vec{v}$.

7.2.4 Total emitted power

We can obtain an expression for the total power emitted by the radiating charge if we evaluate the following integral

$$\int \frac{\mathrm{d}P}{\mathrm{d}\Omega} \frac{\mathrm{d}t}{\mathrm{d}\tau} \mathrm{d}\Omega$$

where all angles are evaluated at time τ

$$\Rightarrow P = \int \frac{q^2 |\vec{g}|^2}{4\pi c^3} \kappa \sin\theta \, \mathrm{d}\theta \, \mathrm{d}\phi.$$

Substituting into (215) and identifying $|\dot{\vec{v}}|\beta \sin i = |\dot{\vec{v}} \times \vec{v}/c|$ and after some manipulation

$$P_{em} = \frac{2e^2}{3c^3} \gamma^6 [|\dot{\vec{v}}|^2 - |\dot{\vec{v}} \times \vec{v}/c|^2]. \tag{217}$$

These equations describe the radiated power of a moving charge. Now we are in a good position to describe specific radiation mechanisms because (217) tells us that the only thing we need to specify is the trajectory of the charge (its velocity and acceleration). As we will see it is the trajectory which defines radiation mechanisms. As examples, I will now describe two of the more important continuum radiation mechanisms in astrophysics, *thermal bremsstrahlung* and *synchrotron* radiation.

7.3 Further reading

Jackson, J.D. (1998) *Classical Electrodynamics*, John Wiley and Sons, New York, NY, USA.

Rybicki, G. B. and Lightman, A. P. (1985) *Radiative Processes in Astrophysics*, John Wiley and Sons, New York, NY, USA.

Shu, F. H. (1992) *The Physics of Astrophysics: Radiation*, University Science Books, Mill Valley, CA, USA.

Tucker, W. H. (1975) *Radiation Processes in Astrophysics*, MIT Press, Cambridge, MA, USA.

Zhelezniakov, V. V. (1996) *Radiation in Astrophysical Plasmas*, Kluwer Academic Publishers, Dordrecht, The Netherlands.

Chapter 8

Bremsstrahlung and synchrotron radiation

8.1 Bremsstrahlung

A thermal plasma in which the ionized particles follow a Maxwell–Boltzmann distribution will undergo Coulomb interactions. The Coulomb forces cause the charged particles to accelerate during the interactions (collisions) so that the particles give off radiation. In this chapter we will study the acceleration of electrons in a thermal plasma and determine the characteristics of the emitted radiation using equation (217).

8.1.1 Single particle collisions

We begin by considering the interaction of a free electron with a free ion, as shown in Fig. 8.1.

$$m_e|\ddot{x}| = \frac{Z_i e^2}{b^2 + v^2 t^2} \quad \text{and} \quad \dot{v} \perp \vec{v} \qquad \frac{v}{c} \ll 1.$$

Substituting this into (217) yields

$$P = \frac{2e^2}{3c^3}|\dot{\vec{v}}|^2 = \frac{2Z_i^2 e^6}{3m_e^2 c^3} \frac{1}{(b^2 + v^2 t^2)^2}. \tag{218}$$

Integrating over the entire collision

$$P_{tot} = \int P(t)\, dt = \frac{2e^2}{3c^3} \int |\dot{\vec{v}}_t|^2 \, dt.$$

This integral has been expressed in the time domain. It can also be expressed in the frequency domain if we recall Parseval's theorem, namely

$$\int_{-\infty}^{\infty} |\dot{\vec{v}}_v|^2 \, dv = \int_{-\infty}^{\infty} |\dot{\vec{v}}_t|^2 \, dt.$$

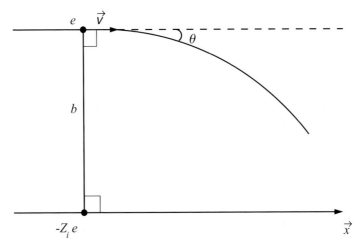

Fig. 8.1 A single particle collision. An electron, of velocity $\vec{v}$, incident on an ion of charge Z is deflected by the Coulomb interaction. The impact parameter is shown as the distance of closest approach (b). The deflection angle is θ.

where

$$|\dot{v}_\nu| = \int |\dot{v}_t| \, e^{2\pi i \nu t} \, dt. \tag{219}$$

Thus

$$\int P_t \, dt = \int P_\nu \, d\nu$$

$$\Rightarrow P_{tot} = \frac{2e^2}{3c^3} \int_{-\infty}^{\infty} |\dot{v}_\nu|^2 \, d\nu = \frac{4e^2}{3c^3} \int_0^{\infty} |\dot{v}_\nu|^2 \, d\nu. \tag{220}$$

The differential form of (220) represents the frequency spectrum resulting from one encounter

$$P_\nu = \frac{4e^2}{3c^3} |\dot{v}_\nu|^2 \, d\nu. \tag{221}$$

Combining (219) and (221)

$$P_\nu = \frac{4e^2}{3c^3} \left[\int |\dot{v}_t|^2 \, e^{2\pi i \nu t} \, dt \right]^2. \tag{222}$$

Combining (218) and (222)

$$P_\nu = \frac{4e^6 Z_i^2}{3m_e^2 c^3} \left[\frac{\pi}{bv} \right]^2 e^{-4\pi |v| b / v}. \tag{223}$$

Equation (223) represents the *single particle spectrum* resulting from a single collision.

8.1.2 Radiation from an ensemble of particles

In reality we must consider the effect of many collisions. Since each collision depends on velocity we must know the velocity distribution of the particles in the plasma. For a thermal plasma that is given by the Maxwell–Boltzmann distribution. What we need to do is to integrate (223) over all velocities and all possible impact parameters. Thus

$$P_v = \int_{v_{min}}^{\infty} [n_e f(v) 4\pi v^2 \, dv] \left[\frac{4\pi^2 Z_i^2 e^6}{3m_e^2 c^3 v^2} \right] \int_{b_{min}}^{\infty} e^{-4\pi |v| b / v} \frac{2\pi b \, db}{b^2}$$

where $f(v)$ is usually a Maxwellian, in which case this simplifies to

$$P_v = n_e \left(\frac{2m_e}{\pi k T} \right)^{1/2} \left(\frac{8\pi^3 Z_i^2 e^6}{3m_e^2 c^3} \right) I \tag{224}$$

where

$$I = \int_{x_{min}}^{\infty} e^{-x} \, dx \int_{\zeta_{min}}^{\infty} \frac{e^{-\zeta}}{\zeta} \, d\zeta$$

where

$$\zeta_{min} \equiv \frac{4\pi v b_{min}}{v} \quad \text{and} \quad x_{min} \equiv \frac{m_e v_{min}^2}{2kT}.$$

For x_{min} we require $(1/2) m_e v_{min}^2 = hv$

$$x_{min} = \frac{hv}{kT}. \tag{225}$$

For b_{min} we require $b_{min} m_e v \approx \hbar$ (QM limit) so that

$$b_{min} = \frac{\hbar}{m_e v} \tag{226}$$

$$\zeta_{min} = \frac{4\pi v \hbar}{m_e v^2}.$$

Combining (224), (225) and (226) yields

$$P_v = \int_{hv/kT}^{\infty} e^{-x} E_1(\zeta_{min}) \, dx. \tag{227}$$

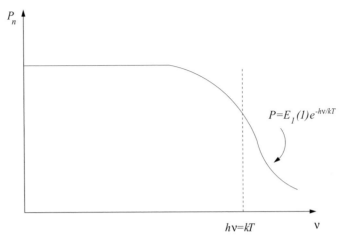

Fig. 8.2 Theoretical bremsstrahlung spectrum. The radiation from a single particle has this spectrum. The fall-off is determined by the thermal energy of the particle.

For

$$\frac{h\nu}{kT} \gg 1 \Rightarrow P_\nu \approx E_1(1)\,e^{-h\nu/kT}.$$

For

$$\frac{h\nu}{kT} \ll 1 \Rightarrow P_\nu \approx \ln\left[\gamma\left(\frac{m_e Z_i^2 e^4}{2\hbar^2 h\nu}\right)\left(\frac{m_e Z_i^2 e^4}{2kT\hbar^2}\right)^{3/2}\right] - \frac{3}{2}\gamma$$

where

$$\gamma = -\int_0^\infty \ln x\,e^{-x}\,dx.$$

The latter is a slowly varying function, therefore the overall spectrum can be approximated according to Fig. 8.2: This, with (227), represent a thermal bremsstrahlung spectrum. These spectra are characteristic of HII regions, supernova remnants and galaxy clusters.

8.2   Synchrotron radiation

In the presence of a magnetic field, charged particles gyrate about the magnetic field. The gyration results in continuous circular acceleration so that the particles radiate. In this section we will describe the total power and frequency spectrum of (a) a single gyrating particle and (b) an ensemble of particles. These results will be applied to astrophysical sources.

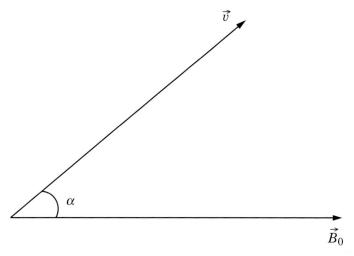

Fig. 8.3 Definition of pitch angle.

8.2.1 Total power

In the presence of a magnetic field a charged particle experiences a Lorentz force

$$\frac{d}{d\tau}(\gamma m \vec{v}) = q\left(\frac{\vec{v}}{c} \times \vec{B}_0\right) \tag{228}$$

Rearranging

$$\frac{d\vec{v}}{d\tau} = \vec{v} \times \hat{z}\omega_B \quad \text{and} \quad \omega_B = \frac{q B_0}{\gamma c} \tag{229}$$

which has the solution

$$\vec{r}(\tau) = \hat{z}v_z\tau + \frac{v_{xy}}{\omega_B}(\hat{x}\cos\omega_B\tau + \hat{y}\sin\omega_B\tau)$$

where

$$v_z = \frac{\vec{v} \cdot \vec{B}_0}{B_0} \quad \text{and} \quad v_{xy} = \sqrt{v_x^2 + v_y^2} = \frac{\vec{v} \times \vec{B}_0}{B_0}$$

for $\vec{B}_0 = B_0\hat{z}$. Letting $\cos\alpha = \vec{v} \cdot \vec{B}_0/v B_0 \Rightarrow \mathcal{R} = (v/\omega_B)\sin\alpha$. The angle α is known as the *pitch angle* and is defined in Fig. 8.3. These descriptions of the particle trajectory can be used to determine the radiation characteristics of the particle through (217). The latter can be cast into a more convenient form

$$P_{em} = \frac{2q^2}{3c^3}\gamma^4(a_\perp^2 + \gamma^2 a_\parallel). \tag{230}$$

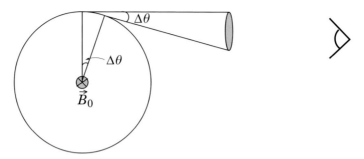

Fig. 8.4 Beamed radiation from a gyrating electron. The electron is moving toward the observer at that particular instant.

Comparing (229) and (230)

$$a_\perp = \omega_B v \sin \alpha \quad \text{and} \quad a_\parallel = 0.$$

Substituting into (230)

$$P_{em} = \frac{2e^2}{3c^3} \gamma^2 \frac{e^2 B_0^2}{m_e^2 c^2} v^2 \sin^2 \alpha \text{ for electrons.}$$

Why can we ignore protons? Because P_{em} scales as $1/m^4$ and protons are 1800 times more massive than electrons. Simplifying this equation

$$\Rightarrow P_{em} = 2\beta^2 \gamma^2 c\sigma_T U_B \sin^2 \alpha$$

where σ_T is the Thomson cross-section and $U_B = B_0^2/(8\pi)$. For an isotropic probability distribution

$$\langle P_{em} \rangle = \frac{4}{3} \beta^2 \gamma^2 c\sigma_T U_B$$

where $\langle \sin^2 \alpha \rangle = 2/3$. For an ensemble of particles $\langle P_{em} \rangle = \langle P \rangle$

$$\Rightarrow \langle P \rangle = \frac{4}{3} \beta^2 \gamma^2 c\sigma_T U_B \tag{231}$$

which describes both emitted and received power.

8.2.2 *The received spectrum*

We will take the approach of examining the properties of the pulses emitted by ultra-relativistic gyrating particles (Fig. 8.4). In terms of the retarded time τ

$$\Rightarrow \frac{\Delta\tau}{P} = \frac{\Delta\theta}{2\pi} \approx \frac{1}{2\pi\gamma}.$$

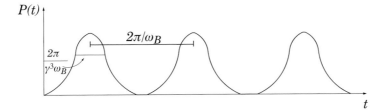

Fig. 8.5 Pulsed radiation from gyrating electron.

In terms of the observer's time, recall

$$\frac{dt}{d\tau} = 1 - \beta\cos\theta \approx 1 - \beta + \frac{(\Delta\theta)^2}{2} \approx \frac{1}{\gamma^2} \quad (\gamma \gg 1)$$

$$\Rightarrow \frac{\Delta t}{P} = \frac{1}{2\pi\gamma^3} \Rightarrow \Delta\nu \approx \gamma^3\omega_B.$$

Using standard pulse analysis we can predict emission at $\omega = \omega_B \to \gamma^3\omega_B$ (Fig. 8.5). Now let us define $\omega_L = \gamma\omega_B = eB_0/m_ec$ and let $\nu = \omega/2\pi$

$$\Rightarrow \langle P_\nu(\gamma)\rangle = \frac{4}{3}\beta^2\gamma^2 c\sigma_T U_B\phi_\nu(\gamma) \tag{232}$$

such that $\int_0^\infty \phi_\nu(\gamma)\,d\nu = 1$. From an ensemble of particles at energy γ we would expect a discrete spectrum with a basic period of

$$\frac{2\pi}{\omega_B} = \frac{\gamma}{\nu_L}.$$

But

$$\nu_L = \frac{eB_0}{2\pi m_ec} \approx 20\text{--}30 \text{ Hz}.$$

In astrophysical sources the basic period is too low to be detectable. In a typical astrophysical spectrum where $\gamma \approx 10^4$

$$\Rightarrow \nu_B = \gamma^{-1}\nu_L = 2 \times 10^{-3} \text{ Hz}$$

the spacing of the harmonics. The spectrum, in practice, is continuous, as shown in Fig. 8.6.

8.2.3 *Spectrum of a power-law energy distribution*

Consider an energy distribution of electrons defined by

$$n(\gamma)\,d\gamma \to m_ec^2\gamma - m_ec^2(\gamma + d\gamma).$$

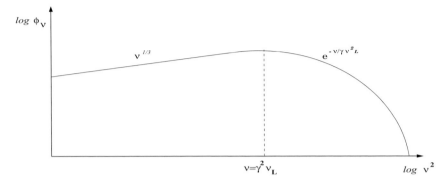

Fig. 8.6 Synchrotron spectrum of a single electron. The gyrating electron emits radiation that is characterized by the spectrum shown. The fall-off frequency is determined by the energy, γ, of the electron.

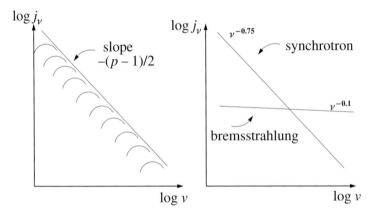

Fig. 8.7 Synchrotron spectrum of an ensemble of electrons. A power-law energy distribution of electrons yields a power-law synchrotron spectrum. The latter can be thought of as the superposition of individual electron spectra, scaled by the electron energy. The slope of the resulting spectrum is shown in terms of the power-law index of the energy distribution. The plot on the right shows typical spectra associated with synchrotron and bremsstrahlung radio sources. The markedly different slopes of the two spectra allow astronomers to differentiate between thermal and non-thermal sources.

With (232) we can define a volume emissivity

$$\rho j_v = \int_1^\infty \langle P_v(\gamma) \rangle n(\gamma) \, d\gamma. \tag{233}$$

Empirical evidence indicates that

$$n(\gamma) \, d\gamma = n_0 \gamma^{-p} \, d\gamma = n_0 \gamma^{-2.5} \, d\gamma. \tag{234}$$

Now substitute (232) and (234) into (233), letting

$$\phi_\nu(\gamma) \rightarrow \delta(\nu - \gamma^2 \nu_L)$$

and let $\beta \approx 1$, $\nu' = \gamma^2 \nu_L$ and integrate to get

$$\rho j_\nu = \frac{2}{3} c \sigma_T n_0 U_B \nu_L^{-1} \left(\frac{\nu}{\nu_L}\right)^{-(p-1)/2}. \tag{235}$$

The basic result is independent of actual shape of ϕ_ν. We can define a luminosity L_ν by integrating (235) over the volume of the source

$$L_\nu = \int_V \rho j_\nu \, dV \propto \nu^{-(p-1)/2}. \tag{236}$$

The net result is a power-law spectrum, as shown in Fig. 8.7.

8.3 Further reading

Jackson, J. D. (1998) *Classical Electrodynamics*, John Wiley and Sons, New York, NY, USA.

Pacholczyk, A. G. (1970) *Radio Astrophysics. Nonthermal Processes in Galactic and Extragalactic Sources*, Freeman, San Francisco, USA.

Rybicki, G. B. and Lightman, A. P. (1985) *Radiative Processes in Astrophysics*, John Wiley and Sons, New York, NY, USA.

Shu, F. H. (1992) *The Physics of Astrophysics: Radiation*, University Science Books, Mill Valley, CA, USA.

Tucker, W. H. (1975) *Radiation Processes in Astrophysics*, MIT Press, Cambridge, MA, USA.

Verschuur, G. A. and Kellerman, K. I. (1988) *Galactic and Extragalactic Radio Astronomy*, Springer-Verlag, Berlin, Germany.

Zhelezniakov, V. V. (1996) *Radiation in Astrophysical Plasmas*, Kluwer Academic Publishers, Dordrecht, The Netherlands.

Chapter 9

High energy processes in astrophysics

Unlike the black-body radiation that characterizes stars and planets, the bremsstrahlung and synchrotron processes are tracers mainly of high-energy phenomena. The former often traces regions of hot ionized gas, the latter traces non-thermal matter consisting of relativistic particles. As we will see later, both processes are associated with supernova remnants and accreting neutron stars. I now examine these phenomena in more detail.

9.1 Neutron stars

When matter accretes onto a neutron star, the energy gained by a particle (assume a proton) is just the potential energy per particle at the neutron star surface

$$U_p = -\frac{GMm_p}{R} \approx 2 \times 10^{-4} \text{ ergs} \approx 0.1 \, m_p c^2.$$

Thus, a thermalized particle at the surface has a temperature of

$$kT \approx 10^8 \text{ eV} \approx 100 \text{ MeV}.$$

In principle particles can radiate γ rays. However, the Eddington limit (more details later) sets a maximum luminosity of

$$L = 4\pi R^2 \sigma T^4 \approx 10^{37} \text{ erg s}^{-1}$$
$$\Rightarrow T \approx 10^7 \text{ K} \Rightarrow kT \approx 1 \text{ keV}.$$

Matter at this temperature is a fully ionized plasma so that collisions are mediated by Coulomb interactions. This, of course, is a recipe for thermal bremsstrahlung.

As noted in the previous chapters, bremsstrahlung is characterized with a high energy cut-off defined by,

$$h\nu \approx kT.$$

144

Thus, for $kT \approx 1$ keV $\rightarrow h\nu \approx 1$ keV. The accreting region should therefore emit to hundreds of electronvolts, consistent with their designation as UV and X-ray sources.

We can contrast this with HII regions ($kT \approx 1$ eV). In this case bremsstrahlung does not operate at the quantum limit and $b_{min} \approx Z_i e^2 / m_e v^2$. Comparing with the quantum limit (189) we see that, in this case, b_{min} is larger by a factor of $Z_i e^2 / \hbar \nu \approx 4 \times 10^8$. The photon energies are therefore smaller by a similar factor, yielding frequencies of order 1 GHz. The HII regions are known to be radio sources.

9.2 Supernova remnants

Supernovas leave behind a neutron star (sometimes a black hole). They also eject a shell of material which is called a supernova remnant (SNR). Detection of 1 keV photons from SNRs suggests that

$$kT \approx 10^3 \text{ eV} \Rightarrow T \approx 10^7 \text{ K.}$$

The bremsstrahlung process operates here too! Since it depends on n_e as well as T the density can be evaluated once T is known. Typical values of n_e are 0.1–10 cm^{-3}, averaged over the SNR.

How much energy does it take to ionize this much gas?
Consider a 10 pc SNR. Then

$$n_e kT V_{snr} \approx (10)(2 \times 10^{-9})(3 \times 10^{19})^3$$
$$\approx 5 \times 10^{50} \text{ ergs} \Rightarrow \text{ blast energy!}$$

Where does the energy come from?

$$\frac{\rho v^2}{2} \approx n_e kT \Rightarrow v \approx \sqrt{\frac{2kT}{m_p}} \approx 5 \times 10^7 \text{ cm s}^{-1}$$
$$\approx 500 \text{ km s}^{-1}.$$

But the sound speed

$$c_s \approx 30 \text{ km s}^{-1} \Rightarrow \mathcal{M} \approx 17 \rightarrow \text{ supersonic.}$$

The supersonic shock from the explosion heats the interior of the SNR.

9.2.1 Particle acceleration

Given the large amount of kinetic energy associated with SNRs, is it possible that SNRs are the source of most of the relativistic particles in the Galaxy? At the Earth we detect these particles as cosmic rays. From all sky radio surveys (see Fig. 8.7) we know that the Galaxy is filled with cosmic rays because the radio emission is almost entirely synchrotron.

$$\text{Recall} \qquad \nu_c = \frac{eB}{2\pi m_e c}\gamma^2.$$

For $B = 5\mu G$, $\gamma = 2 \times 10^4 \Rightarrow \nu = 3$ GHz. At radio frequencies, synchrotron-emitting electrons are highly relativistic.

Observations of synchrotron emission in SNRs indicates that relativistic particles are also present there. For SNRs to be the primary source of cosmic rays we need a mechanism that taps the blast energy of the SNR. That mechanism is known as *diffusive shock acceleration* (see Gaisser, 1990).

To determine whether SNRs are plausible sources of cosmic rays we need to evaluate how much of an SNR's energy is in the form of cosmic rays.

Minimum energy

Consider an SNR of volume V, containing a mixture of cosmic rays and magnetic fields. The energy density of the magnetic field is simply

$$U_B = \frac{B_0^2}{8\pi}. \tag{237}$$

The energy density of the cosmic rays is a little more difficult to determine. For electrons

$$U_e = n_0 \int_{\gamma_{min}}^{\infty} (\gamma m_e c^2)\gamma^{-p}\,d\gamma = \frac{n_0 m_e c^2}{2-p}\gamma_{min}^{-(p-2)} \qquad [p>2] \tag{238}$$

requiring knowledge of the electron energy density distribution. But, from synchrotron theory, $P_\nu \propto n_0 B_0^{-(p+1)/2}$, so that

$$n_0 = \text{constant} \times B_0^{(p+1)/2} \tag{239}$$

at a fixed frequency, ν, of observation.

Combining (238) and (239)

$$\Rightarrow U_e = \text{constant}\, B_0^{-(p+1)/2} \text{ at fixed } \nu \text{ (fixed } \gamma\text{)}. \tag{240}$$

The total non-thermal energy can now be expressed in terms of the magnetic field strength

$$U = U_B + kU_e = U_B + U_{CR}$$
$$\Rightarrow U \propto B_0^2 + B_0^{-(p+1)/2} \tag{241}$$

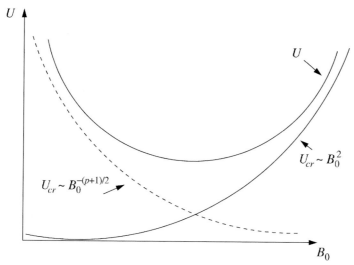

Fig. 9.1 Minimum energy. The energy in particles and fields is plotted as a function of B_0. The curve has a well-defined minimum.

from (237) and (240) and where $k = U_{CR}/U_e$ (which can be estimated from observations of cosmic rays at the Earth). A plot of (241) is shown in Fig. 9.1. Equation (241) has a minimum when

$$\frac{d}{dB_0}(U_{CR} + U_B) = -\frac{(p+1)U_{CR}}{2B_0} + \frac{2U_B}{B_0} = 0$$

$$\Rightarrow U_{CR} = \frac{4}{p+1}U_B.$$

For $p \approx 2.5$, $U_{CR} \approx U_B$. There is an approximate balance between the cosmic ray and magnetic field energy densities. This balance is known as *equipartition*.

From this we see that the minimum energy can now be expressed as

$$U_{min} = (U_{CR} + U_B)_{min} \approx 2U_{CR} \approx 2U_B.$$

The equipartition condition allows the calculation of B_0 which yields U_{CR}, U_B and U_{min}. Regardless of assumptions about equipartition, U_{min} can always be calculated. The non-thermal energy curve for a SNR in the nearby galaxy, M33, is shown in Fig. 9.2. Arguments based on equipartition have been used to estimate how much cosmic ray energy resides inside SNRs. These calculations suggest

$$\langle U_{min} \rangle \approx 5 \times 10^{49} \text{ ergs}$$

which is a substantial fraction of the blast energy (estimated to be $\approx 10^{51}$ ergs). On the basis of these numbers it seems plausible that SNRs convert a fair fraction

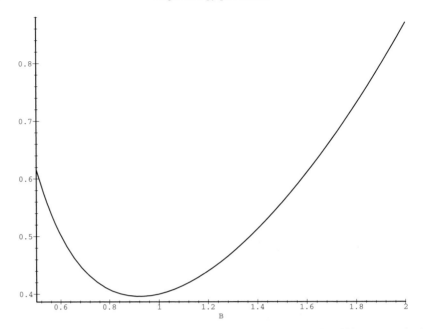

Fig. 9.2 The minimum energy curve for a supernova remnant in M33. The vertical axis has units of 10^{50} ergs. The horizontal axis has units of 10^{-4} gauss.

of their blast energy into cosmic rays (Fig. 9.3). The SNRs are therefore a leading contender as the source of most cosmic rays with energies up to 10^{14} eV. A VLA image of the SNR Cas A is shown in Fig. 9.4. The emission is dominated by synchrotron radiation indicating the presence of cosmic ray electrons. Cosmic rays are found throughout the Galaxy, as evidenced by the synchrotron emission from the Milky Way plane (Fig. 9.5).

9.3 Radio galaxies

The center of our Galaxy is a powerful source of synchrotron radio emission (Fig. 9.6). There is a debate as to whether the center is powered by a massive black hole ($10^5 M_\odot$) or by a concentrated cluster of stars. Although a formidable source of energy, the center of our Galaxy pales in comparison with those of *active galaxies*.

Perhaps the most powerful objects in the Universe, radio galaxies, are highly active galaxies containing large reservoirs of non-thermal energy

$$U_{min} \approx 10^{60} \text{ ergs} > 10^9 SNe$$

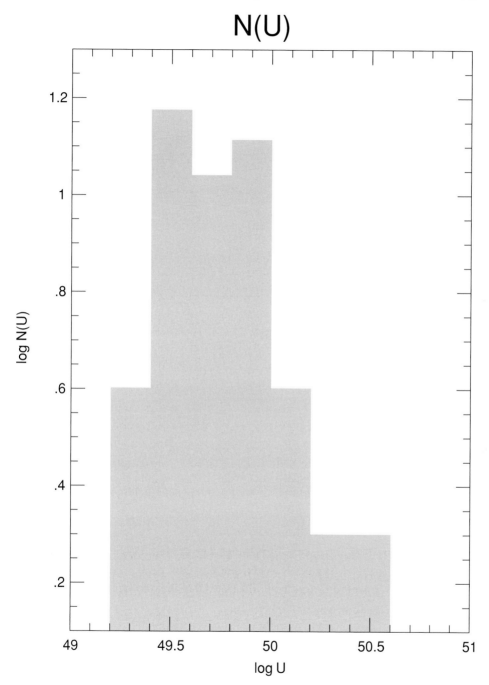

Fig. 9.3 A histogram of minimum energies calculated for a sample of about 50 SNRs in M33. The energy is in ergs. The sample is complete above log $U \approx 49.5$.

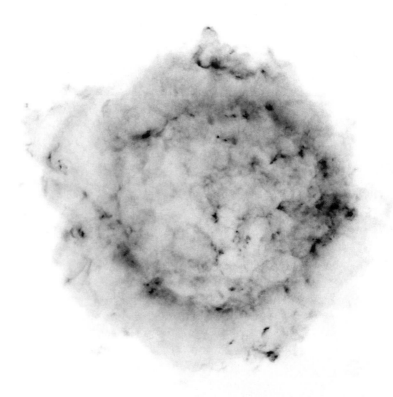

Fig. 9.4 A VLA image of the SNR, Cas A. The observing frequency is 1.4 GHz. The radio emission is dominated by synchrotron radiation indicating that the SNR contains relativistic electrons gyrating in magnetic fields. Image courtesy of NRAO/AUI.

located hundreds of kiloparsecs from the power centers of these galaxies, their nuclei. A VLA image of Cygnus A, a well-known radio galaxy, is shown in Fig. 9.7.

A simple light travel time argument illustrates that the relativistic particles filling these reservoirs must be accelerated in situ.

The lifetime of a relativistic particle is ultimately dependent on the rate at which it loses energy (emits synchrotron radiation).

$$m_e c^2 \frac{\mathrm{d}\gamma}{\mathrm{d}\tau} = -P_{em} = -2\beta^2 \gamma^2 c \sigma_T U_B \sin^2 \alpha$$

$$\Rightarrow -\frac{\gamma}{\mathrm{d}\gamma/\mathrm{d}\tau} = \frac{m_e c/\gamma}{2\sigma_T U_B \sin^2 \alpha}.$$

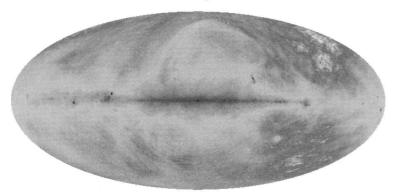

Fig. 9.5 The radio sky at 408 MHz. The grey-scale represents intensity of radio emission from the sky with black being most intense. The Galactic plane is horizontal in this representation. Although radio emission is evident across the whole sky there is a concentration of emission toward the plane and the center of the Galaxy. From: http://www.mpe.mpg.de/~hcs/Cen-A/Pictures/galaxy-radio408mhz-2.gif.

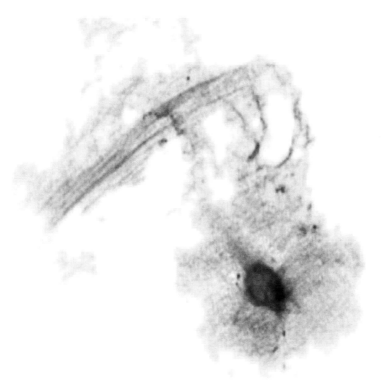

Fig. 9.6 A VLA image of the Galactic center. The observing frequency is 1.4 GHz. The filamentary structures are not well understood but are believed to be related to violent activity at the Galactic center. Image courtesy of NRAO/AUI.

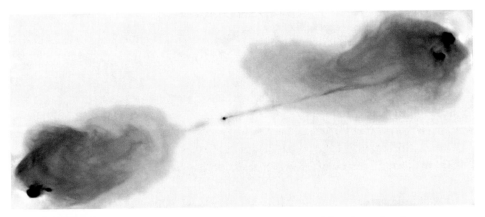

Fig. 9.7 A VLA image of the active galaxy Cygnus A. The observing frequency is 1.4 GHz. The two outer lobes are intense sources of synchrotron emission suggesting that the lobes are filled with relativistic electrons and magnetic fields. The nucleus of Cen A is the relatively faint point source at the center of the image. Image courtesy of NRAO/AUI (Perley, R. C., Carilli, C. and Dreher, J.).

Typically, $B = 10^{-5}$ G, $\gamma \approx 10^4 \rightarrow \tau_e \approx 10^7$ years. For $\gamma = 10^6 \rightarrow \tau_e \approx 10^5$ years. But the light-travel-time for a distance of $l \approx 10^{25}$ cm $\Rightarrow \tau_c \approx l/c \approx 10^7$ years. So, how is energy transferred in radio galaxies? The particles must be accelerated within the lobes and perhaps all along the way from the nucleus to the lobes. If that is the case, there is no need for the particles to be completely transferred to the lobes. The acceleration occurs in situ.

9.4   Galactic X-ray sources

As we have discussed earlier, rotation powers some objects such as pulsars. In binary stars, infalling matter can also power neutron stars and white dwarfs. The latter are called *accretion powered sources*. They are among the most powerful sources in the Galaxy and emit much of their radiation in the high-energy part of the EM spectrum. They were discovered by the earliest X-ray telescopes (such as Uhuru) and studied in great detail by subsequent X-ray and γ-ray satellites such as Copernicus, Ariel 5, SAS-3, OSO-7.8, COS-B, HEAO-1, Einstein, EXOSAT, ROSAT, ASCA, XTE and Chandra. These X-ray sources are characterized by X-ray luminosities of L_x (2–10 keV) $\approx 10^{35}$–10^{37} erg s^{-1}. Since they are always found in binary star systems the source of the energy is believed to be mass transfer from a companion star to the compact star (white dwarf, neutron star, black hole) hence the name *X-ray binaries*.

9.4.1 The energy source

Matter falling onto a compact star represents the conversion of gravitational potential energy into radiation. Thus, as discussed earlier,

$$L_x \approx \dot{M}\frac{GM_c}{R_c} = \dot{M}c^2\left(\frac{GM_c}{c^2R_c}\right) = \dot{M}c^2\left(\frac{2R_s}{R_c}\right)$$

where, M_c, R_c are the mass and radius of the compact object and R_s is the Schwarzschild radius of the compact object. For a white dwarf, $M_c \approx 1M_\odot$, $R_c \approx 1\,000$ km. The efficiency of energy conversion is given by

$$\approx \left(\frac{GM_c}{c^2R_c}\right) \approx 0.001$$

or about 0.1%. For neutron stars, $M_c \approx 1M_\odot$, $R_c \approx 10$ km

$$\Rightarrow \text{ efficiency} \approx 0.1$$

or about 10%. For a black hole the efficiency is also about 10%. These numbers contrast with 0.7% for thermonuclear burning of hydrogen and 0.1% for burning of heavier elements. Neutron stars and black holes are the most efficient astrophysical engines known. In order for these engines to produce the observed X-ray luminosities of $L_x \approx 10^{37}$ erg s^{-1} the required rate of mass accretion is $\dot{M} \approx 10^{-9}M_\odot$/year for a neutron star and $\dot{M} \approx 10^{-7}M_\odot$/year for a white dwarf.

9.4.2 Maximum luminosity/Eddington limit

It is no accident that X-ray binaries have a well-defined upper limit to their luminosities. As compact stars accrete more matter they become more luminous. The increased radiation pressure opposes the gravitational infall of matter and eventually an equilibrium is reached at the so-called Eddington luminosity. This can be investigated quantitatively. We begin with the simplifying assumption that the infalling matter is pure, fully ionized hydrogen. Then

$$F_g = -\frac{GM_c\rho}{r^2} = -\frac{GM_c nm_p}{r^2}$$

$$F_r = \frac{L}{4\pi r^2 c}n\sigma_T \qquad \text{spherical symmetry}$$

where $\sigma_T = 6.6 \times 10^{-25}$ cm^2 is the Thomson cross-section and F_g and F_r are the gravitational and radiation forces respectively. An equilibrium is reached when

$$F_r = -F_g \Rightarrow L = L_{Edd} = \frac{4\pi GM_c c m_p}{\sigma_T} = 1.3 \times 10^{38}\left(\frac{M_c}{M_\odot}\right) \text{ erg s}^{-1}.$$

9.4.3 Characteristic temperature

What would the surface temperature of the compact star have to be in order to thermally emit the required 10^{37} erg s^{-1}? With

$$L = 4\pi R_c^2 \sigma T_c^4$$

and $\sigma = 5.7 \times 10^{-5}$ erg cm^{-2} degrees^{-4} s^{-1}

$$T \approx 10^7 \left(\frac{R}{10\text{ km}}\right)^{-1/2} \quad \text{K} \rightarrow kT \approx \left(\frac{R}{10\text{ km}}\right)^{-1/2} \text{keV}.$$

Neutron stars and black holes are *natural* X-ray emitters. White dwarfs are *natural* UV sources. The emission we have discussed thus far can be characterized as black body and originates in the compact accretion disks that form around the compact star (see Section 9.4.4 next). The black-body radiation is optically thick bremsstrahlung. These sources also emit optically thin bremsstrahlung and synchrotron radiation.

9.4.4 Mass transfer

We now investigate how matter is transferred onto the compact object.

Roche Lobe overflow

In Fig. 9.8, matter from one star *spills over* the equipotential surfaces that define the potential of the two-body system.

Stellar winds

A variant of the above picture suggests that the mass-carrying wind of the evolving star is *focused* onto the compact object by the latter's intense gravitational field (Fig. 9.9). Upper main-sequence stars have

$$\dot{M}_{wind} \approx 10^{-7} - 10^{-6} M_\odot / \text{ year}$$
$$\Rightarrow \dot{M}_{acc} \approx 0.1\% \text{ of } \dot{M}_{wind}$$
$$\Rightarrow \dot{M}_{acc} \approx 10^{-9} - 10^{-10} M_\odot / \text{year, which is adequate.}$$

9.5 Accretion disks

Matter falling onto the compact star forms an accretion disk. How? The observational evidence for white dwarfs comes in the form of cataclysmic variables where we see Doppler shifts from gas corresponding to Keplerian velocities at the Roche Lobe. In the case of neutron stars we observe X-ray binaries. The best case is Hercules X-1 which shows evidence for a large precessing disk. A schematic representation of the evolution of an accretion disk is shown in Fig. 9.10.

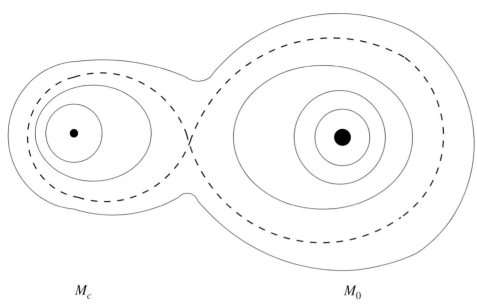

Fig. 9.8 Potential surfaces of a binary star. The compact star is shown as M_c. Gas spills over from M_0 to M_c.

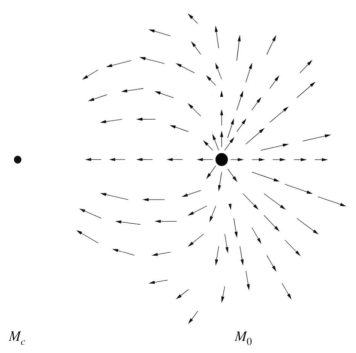

Fig. 9.9 Focused wind in a binary star system. The stellar wind of the evolving star is focused by the gravitational field of the compact star, resulting in mass accretion.

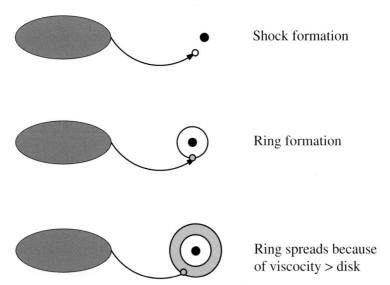

Fig. 9.10 Formation of an accretion disk. The formation of an accretion disk occurs in stages, beginning with the formation of a shock as the material piles up during infall, and ends with ring spreading into a disk because of gas viscosity.

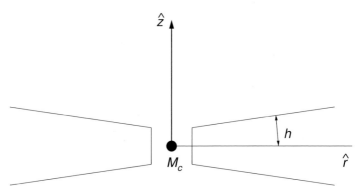

Fig. 9.11 Cross-section of an accretion disk. The parameters used to define the vertical structure of the disk are shown.

9.5.1 *Disk hydrodynamics*

Let us consider a steady-state axisymmetric thin disk (Fig. 9.11). We will briefly discuss the vertical structure of the disk. A description of the full structure of the disk requires consideration of the equations of *mass conservation*, *conservation of radial momentum*, *conservation of angular momentum* and the *conservation of energy*, which is beyond the scope of this book.

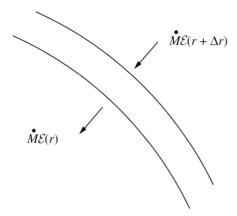

Fig. 9.12 Pressure gradients in an accretion disk.

Vertical structure

The vertical (z direction) scale length is defined as

$$h \equiv \frac{1}{\rho_0} \int_0^\infty \rho(z) \, \mathrm{d}z.$$

In order for the accretion disk to be in hydrostatic equilibrium

$$\frac{\mathrm{d}P}{\mathrm{d}z} = -\rho g_z = -\rho(z) \frac{GM}{r^2} \frac{z}{r}$$

$$P = \frac{GM}{r^3} \int_0^\infty \rho(z) z \, \mathrm{d}z \approx \frac{GM\rho h^2}{2r^3}$$

$$\left(\frac{h}{r}\right)^2 = 2 \frac{P/\rho}{GM/r} \approx 2 \frac{c_s^2}{V_P} \approx \frac{c_s^2}{V_k}.$$

Since $c_s \ll V_k$, we see that $h/r \ll 1$. The accretion disk is very thin. Consequently

$$\frac{P}{r} \ll \frac{GM\rho}{r^2}.$$

The pressure gradients are very small in the radial direction (Fig. 9.12).

9.5.2 The emission spectrum of the disk

In order to determine the emission spectrum of the accretion disk we begin by crudely estimating the rate of energy deposition per unit area, Q (Fig. 9.13).

$$Q2\pi r \Delta r = \dot{M} \left[-\frac{GM}{2(r + \Delta r)} + \frac{GM}{2r} \right] = \frac{GM\dot{M}}{2r^2} \Delta r$$

$$\Rightarrow Q = \frac{GM\dot{M}}{4\pi r^3}.$$

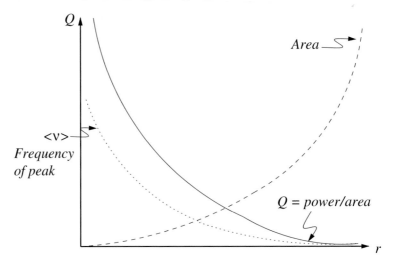

Fig. 9.13 Functional dependence of Q, peak frequency and the area.

When viscosity is included (a form of energy transport)

$$Q = \frac{3}{8\pi} \frac{GM\dot{M}}{r^3}.$$

The spectral signature of the accretion disk is the empirically determined $L_\nu \propto \nu^{1/3}$ law. Let us see if we can model such a law theoretically. From the previous discussion

$$Q(r) \propto \frac{1}{r^3}.$$

Thus, for black-body emission

$$Q = \sigma T^4 \propto \frac{1}{r^3} \rightarrow T \propto r^{-3/4}$$

and

$$\nu_{BB} \approx kT \propto r^{-3/4}$$

$$\Rightarrow L_\nu = 2\pi \int r \, dr \, S(\nu, r)$$

$$\int_0^\infty S(\nu, r) \, d\nu = Q(r) = \frac{Q_0}{r^3}.$$

Let us treat $S(\nu, r)$ as a sharp function of frequency such that

$$S(\nu, r) \approx \frac{Q_0}{r^3} \delta \left[\nu - \nu_0 \left(\frac{r}{r_0} \right)^{-3/4} \right].$$

We are using a similar approach to that we used when discussing synchrotron emission.

Now, let us define

$$\delta\left[\nu - \nu_0 \left(\frac{r}{r_0}\right)^{-3/4}\right] \equiv \delta[f(\nu, r)]$$

so that

$$L_\nu \approx 2\pi \int_r dr \frac{Q_0}{r^2} \delta[f(\nu, r)]$$

$$= 2\pi Q_0 \int \frac{1}{r^2 (df/dr)} \delta(f) \left(\frac{df}{dr}\right) dr$$

$$= 2\pi Q_0 \int \frac{1}{r^2 (df/dr)} \delta(f) df.$$

Now, from this

$$\frac{df}{dr} = \frac{3}{4} \nu_0 r_0^{3/4} r^{-7/4} \propto r^{-7/4}$$

so that

$$L_\nu \propto \left(\frac{r^{7/4}}{r^2}\right)_{f=0} = (r^{-1/4})_{f=0}.$$

Since $f = \nu - \nu_0 (r/r_0)^{-3/4} = 0$

$$\Rightarrow r^{-3/4} \propto \nu \rightarrow r^{-1/4} \propto \nu^{1/3}.$$

So, finally

$$L_\nu \propto \nu^{1/3}.$$

The accretion disks around neutron stars are characterized by this spectrum which incorporates black-body radiation (at low frequencies), a $\nu^{1/3}$ law at intermediate frequencies and a high-frequency bremsstrahlung fall-off (Fig. 9.14).

9.6 Pulsars revisited

In addition to being astrophysically interesting, pulsars also represent interesting applications of what we have learned in previous chapters. The reason for the latter is that they have both static and radiation fields with a well-defined transition region separating the two EM regimes.

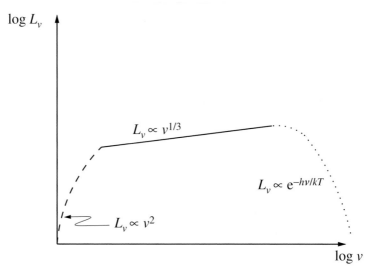

Fig. 9.14 Emission spectrum of an accretion disk. The optically thick regime is shown as a dashed curve at low frequencies. The optically thin regime is shown as a solid line while the fall-off is shown as a dotted line.

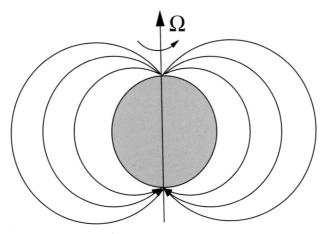

Fig. 9.15 Aligned magnetic field. The B field is said to be aligned with the rotation axis of the pulsar.

9.6.1 The radiation field

Neutron stars have intense dipolar magnetic fields which are anchored to the star and therefore rotate with it (Fig. 9.15). The rotating magnetic field is responsible for generating a strong static electric field as well as radiation. To see that a rotating magnetic field does radiate let us consider its rotation.

$$B = B_0 \left(\frac{r}{R_s} \right)^{-3}.$$

The rapid rotation of the B field ensures that at some radius, R_L, the linear velocity of the B field must be that of light, that is when

$$\Omega = \frac{c}{R_L}.$$

Thus, inside R_L we have $v = r\Omega < R_L\Omega = c$ and the fields are said to *co-rotate* with the star. On the other hand, outside R_L, $v = r\Omega > R_L\Omega > c$ and co-rotation is not possible. For $r > R_L$ the rotating B field must radiate away energy in order not to violate causality.

9.6.2 *Radiated power*

We can get an estimate of how much power must be radiated by examining the energy density of the magnetic field at the *light cylinder*.

$$U \approx \frac{B^2}{8\pi} = \frac{1}{8\pi} \left[B_s \left(\frac{R_L}{R_s} \right)^{-3} \right]^2 = \frac{B_s^2 \, R_s^6}{8\pi \, R_L^6}.$$

From the definition of the Poynting vector

$$S = Uc$$

$$\Rightarrow \dot{E} \approx -4\pi R_L^2 Uc = \frac{1}{2} B_s^2 R_s^6 R_L^{-4} c$$

$$= -\frac{1}{2} B_s^2 R_s^6 \Omega^4 c^{-3}.$$

Thus

$$\frac{\mathrm{d}}{\mathrm{d}t} \left(\frac{I\Omega^2}{2} \right) = I\Omega\dot{\Omega} = -\frac{B_s^2 R_s^6 \Omega^4}{2c^3}.$$

So that,

$$\dot{\Omega} \propto -\Omega^3$$

$$\Omega \propto t^{-1/2} \quad \text{if} \quad \Omega \ll \Omega_{\text{initial}}$$

$$P \propto t^{1/2}$$

$$\Rightarrow t = \frac{P}{2\dot{P}}.$$

For the Crab pulsar $t = 1300$ years. The predicted age of the Crab pulsar is not far from its known age (we know the Crab supernova went off in AD 1054, almost 1000 years ago).

9.6.3 The Braking Index

It is conventional to characterize the spindown of a pulsar by the so-called *Braking Index*, n defined by

$$\dot{\Omega} = -k\Omega^n$$

or

$$\frac{\ddot{\Omega}\Omega}{\dot{\Omega}^2} = n.$$

For magnetic dipole radiation $n = 3$.

9.6.4 The static magnetic field

We can use these equations to estimate the magnetic field strength of the Crab pulsar.

$$B^2 = -\frac{I\dot{\Omega}c^3}{R_s^6\Omega^3} = \frac{I}{(2\pi)^2}\frac{P\dot{P}c^3}{R_s^6}$$

$$B \approx (P\dot{P})^{1/2}\frac{I^{1/2}c^{3/2}}{2\pi R_s^3} \approx 3.3 \times 10^{19}(P\dot{P})^{1/2}\left(\frac{I}{10^{45}\text{ g cm}^{-3}}\right)^{1/2}\left(\frac{R}{10\text{ km}}\right)^{-3}.$$

For the Crab pulsar, for which P and $\dot{P}$ have been accurately measured,

$$B \approx 3 \times 10^{12}\text{ G}.$$

The range for typical pulsars is given by

$$10^{11} < B < 10^{13}\text{ G}.$$

9.6.5 The static electric field

A rotating magnetic field (Fig. 9.16) will generate an electric field according to

$$\vec{E} = \frac{\vec{v}}{c} \times \vec{B}$$

where

$$|\vec{v}| = R\Omega.$$

A charged particle will accelerate along the E field and gain energy before leaving the pulsar magnetosphere. The amount of energy gain for a particle of charge q is

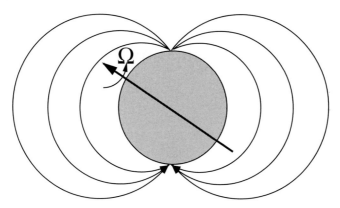

Fig. 9.16 Misaligned magnetic field. The axis of the B field is not aligned with the axis of rotation.

given by

$$\mathcal{E} = eEl \approx \frac{e\Omega R}{c} Bl$$

$$= \frac{e2\pi RBl}{Pc} = 63 \text{ GeV} \left(\frac{R}{10 \text{ km}}\right) \left(\frac{B}{10^{12}\,\text{G}}\right) \left(\frac{l}{\text{cm}}\right)$$

where l is the distance over which the particle is accelerated. For $l \approx R$, $\mathcal{E} = 6.3 \times 10^{16}$ eV. Thus, in principle, the electric field of a pulsar is so strong that it can accelerate particles to almost 10^{17} eV. Such particles form the highest energy *cosmic rays*. The origin of cosmic rays, in general, and the origin of the highest energy cosmic rays, in particular, is a hot topic in astrophysics in 2003 and has yet to be resolved.

9.7 Reference

Gaisser, T. K. (1990) *Cosmic Rays and Particle Physics*, Cambridge University Press, Cambridge, UK.

9.8 Further reading

Clayton, D. D. (1998) *Principles of Stellar Evolution and Nucleosynthesis*, University of Chicago Press, Chicago, USA.

Frank, J., King, A., Raine, D. (2002) *Accretion Power in Astrophysics*, Cambridge University Press, Cambridge, UK.

Rybicki, G. B. and Lightman, A. P. (1985) *Radiative Processes in Astrophysics*, John Wiley and Sons, New York, NY, USA.

Shapiro, S. L. and Teukolsky, S. A. (1983) *Black Holes, White Dwarfs and Neutron Stars*, John Wiley and Sons, New York, USA.

Shu, F. H. (1992) *The Physics of Astrophysics: Radiation*, University Science Books, Mill Valley, CA, USA.

Chapter 10

Electromagnetic wave propagation

So far we have assumed that the radiation we receive from astronomical sources is unaltered en route to the observer. However, in some cases this assumption is not correct. I will now examine the dispersion of EM radiation and the related phenomenon of Faraday rotation, both of which are important at radio frequencies. I begin with Maxwell's Equations,

$$\vec{\nabla} \cdot \vec{E} = 4\pi \rho_e \tag{242}$$

$$\vec{\nabla} \times \vec{E} = -\frac{1}{c}\frac{\partial \vec{B}}{\partial t} \tag{243}$$

$$\vec{\nabla} \times \vec{B} = \frac{4\pi}{c}\vec{J} + \frac{1}{c}\frac{\partial \vec{E}}{\partial t} = -\frac{4\pi}{c}\rho_e\vec{V} + \frac{1}{c}\frac{\partial \vec{E}}{\partial t} \tag{244}$$

$$\vec{\nabla} \cdot \vec{B} = 0. \tag{245}$$

Consider also the force law for charged particles

$$\rho_m \frac{\partial \vec{V}}{\partial t} = -\rho_e \vec{E} - \frac{\rho_e}{c}(\vec{V} \times \vec{B}). \tag{246}$$

These equations describe the interplay between plasmas and EM waves.

Consider a plasma with no large scale $\vec{E}$ and $\vec{B}$ fields and an EM wave passing through it. The plasma sees oscillating $\vec{E}$ and $\vec{B}$ fields

$$\vec{E}_1 = \vec{E}_{10}\, e^{i(\vec{k}\cdot\vec{r}-\omega t)} \qquad \vec{B}_1 = \vec{B}_{10}\, e^{i(\vec{k}\cdot\vec{r}-\omega t)}. \tag{247}$$

The electrons in the plasma, being lighter, oscillate in response. We therefore look for electron velocity solutions of the form

$$\vec{V} = \vec{V}_0\, e^{-i\omega t}. \tag{248}$$

10.1   EM waves in an un-magnetized plasma

When $\vec{B} = 0$,

$$\Rightarrow \vec{V} \times \vec{B} = 0 \quad \text{(in (246))}.$$

Combine $\vec{\nabla} \times$ (243) and $\partial/\partial t$ (244) and (246), and eliminate $\partial \vec{V_e}/\partial t$ and $\vec{B}$

$$\Rightarrow -c\vec{\nabla} \times (\vec{\nabla} \times \vec{E}) = \frac{4\pi \rho_e^2}{\rho_m c}\vec{E} + \frac{1}{c}\frac{\partial^2 \vec{E}}{\partial t^2}.$$

Use identity $\vec{\nabla} \times (\vec{\nabla} \times \vec{E}) = \vec{\nabla}(\vec{\nabla} \cdot \vec{E}) - \nabla^2 \vec{E}$.

Define $\omega_e^2 = 4\pi \rho_e^2/\rho_m = 4\pi n_e e^2/m_e$ – Plasma Frequency

$$\Rightarrow c\nabla^2 \vec{E} = \frac{\omega_e^2}{c}\vec{E} + \frac{1}{c}\frac{\partial^2 \vec{E}}{\partial t^2}. \tag{249}$$

Consider the 1-D case

$$c\frac{\partial^2 E}{\partial x^2} = \frac{\omega_e^2}{c}E + \frac{1}{c}\frac{\partial^2 E}{\partial t^2}. \tag{250}$$

Look for plane wave solutions described by (247)

$$-ck^2 E = \frac{\omega_e^2}{c}E - \frac{\omega^2}{c}E$$

$$\Rightarrow E\left(\frac{\omega^2}{c} - ck^2 - \frac{\omega_e^2}{c}\right) = 0$$

$$\Rightarrow \omega^2 = \omega_e^2 + c^2 k^2$$

which is the *Dispersion Relation*. We can now define the index of refraction, the phase speed and the group velocity.

Index of Refraction

$$\eta = \frac{ck}{\omega} = \frac{\sqrt{\omega^2 - \omega_e^2}}{\omega} = \sqrt{1 - \frac{\omega_e^2}{\omega^2}}$$

$\omega_e < \omega \rightarrow 0 < \eta < 1$

$\omega_e = \omega \rightarrow \eta = 0$

$\omega_e > \omega \rightarrow \eta = $ imaginary $\rightarrow$ reflection and absorption

Phase Speed

$$V_p = \frac{\omega}{k} = \frac{1}{k}\sqrt{\omega_e^2 + k^2 c^2} \tag{251}$$

$\omega_e > 0 \rightarrow V_p > c$

$\omega_e = 0 \rightarrow V_p = c$ (vacuum)

$\omega_e \gg kc \rightarrow V_p = \omega_e/k$

Group Velocity

$$V_g = \frac{\partial \omega}{\partial k} = \frac{c^2 k}{\sqrt{\omega_e^2 + c^2 k^2}} \tag{252}$$

$$\omega_e > 0 \rightarrow V_g < c$$

$$\omega_e = 0 \rightarrow V_g = c$$

$$\omega_e \gg ck \rightarrow V_g = kc^2/\omega_e \rightarrow 0.$$

10.1.1 Dispersion measure

To relate these to astronomical phenomena, let us consider observations of radio frequency signals emitted by pulsars. A pulse is a wave packet and its propagation is therefore characterized by its group velocity. Dispersion has the effect of spreading the pulse such that different pulse frequencies will arrive at different times. Consider a pair of observations made at frequencies ω_1 and ω_2. The arrival times are therefore t_1 and t_2 and are governed by the phase velocity. The difference in arrival times is

$$\Delta t = t_1 - t_2 = \frac{l}{V_1} - \frac{l}{V_2} \tag{253}$$

where V_1 and V_2 are the phase velocities at ω_1 and ω_2 and l is the distance to the source. Rearranging (251)

$$V_p = \frac{c}{\sqrt{1 - (\omega_e^2/\omega^2)}}. \tag{254}$$

Normally, $\omega \gg \omega_e$ so that (254) becomes

$$V_p \approx \frac{c}{1 - \frac{1}{2}(\omega_e^2/\omega^2)}. \tag{255}$$

Inserting (255) into (253), cross multiplying and neglecting high order terms in the denominator, yields

$$\Delta t \approx \frac{l\omega_e^2}{2c}\left[\frac{1}{\omega_1^2} - \frac{1}{\omega_2^2}\right] \tag{256}$$

$$\Rightarrow \Delta t \approx \frac{2\pi e^2}{m_e c}\left[\frac{1}{\omega_1^2} - \frac{1}{\omega_2^2}\right]n_e l = 1.35 \times 10^{-3}\left[\frac{1}{\nu_1^2} - \frac{1}{\nu_2^2}\right]\text{DM seconds.} \tag{257}$$

The quantity $n_e l$ is referred to as the dispersion measure and is often denoted as DM. Since n_e is normally a function of l (257) is generalized so that

$$\text{DM} = \int_0^l n_e\, dl' \text{ cm}^{-2}. \tag{258}$$

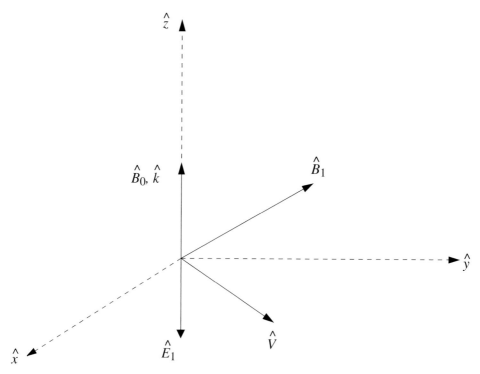

Fig. 10.1 Definition of coordinate system for EM wave propagating in a magnetized plasma.

Application of (255) to observations of pulsars has allowed astronomers to determine the electron density of the ISM in the direction of the pulsars. Typical values are found to be $\approx 0.03 \text{ cm}^{-3}$.

10.2   EM waves in a magnetized medium

Consider an EM wave traveling along a magnetic field line, $\vec{B}_0$, as shown in Fig. 10.1. Treat EM wave fields as perturbations on the plasma fields

$$\vec{E} = \vec{E}_0 + \vec{E}_1 \qquad \vec{B} = \vec{B}_0 + \vec{B}_1.$$

Substitute into (243), (244) and (246) and retain terms to 1st order

$$i\vec{k} \times \vec{E}_1 = \frac{i\omega}{c}\vec{B}_1$$

$$i\vec{k} \times \vec{B}_1 = -\frac{4\pi n_0 e}{c}\vec{V} - \frac{i\omega}{c}\vec{E}_1$$

$$-i\omega m_e \vec{V} = -e\vec{E}_1 - \frac{e}{c}\vec{V} \times \vec{B}_0.$$

A consistent solution must have $\vec{V}_1$, $\vec{E}_1$, $\vec{B}_1$ in the x–y plane.

Let
$$\vec{k} = k\hat{z} \qquad \vec{E}_1 = E_x\hat{x} + E_y\hat{y} \qquad \vec{B}_1 = B_x\hat{x} + B_y\hat{y} \qquad \vec{V} = V_x\hat{x} + V_y\hat{y}.$$

Insert into the above to get

$$-ikE_y = \frac{i\omega}{c}B_x \tag{259}$$

$$ikE_x = \frac{i\omega}{c}B_y \tag{260}$$

$$-ikB_y = -\frac{4\pi}{c}n_0 e V_x - \frac{i\omega}{c}E_x \tag{261}$$

$$ikB_x = -\frac{4\pi}{c}n_0 e V_y - \frac{i\omega}{c}E_y \tag{262}$$

$$-i\omega m_e V_x = -eE_x - \frac{e}{c}B_0 V_y \tag{263}$$

$$-i\omega m_e V_y = -eE_y + \frac{e}{c}B_0 V_x. \tag{264}$$

Insert (259) and (260) into (261) and (262) to eliminate B_x, B_y

$$V_x = \frac{(i\omega/c) - (ik^2 c/\omega)}{-4\pi n_0 e/c}E_x \tag{265}$$

$$V_y = \frac{(i\omega/c) - (ik^2 c/\omega)}{-4\pi n_0 e/c}E_y. \tag{266}$$

Insert (265) and (266) into (263) and (264) to get

$$\begin{pmatrix} i\omega m_e \dfrac{((i\omega/c) - (ik^2 c/\omega))}{4\pi n_0 e/c} + e & \dfrac{eB_0}{c}\dfrac{((ik^2 c/\omega) - (i\omega/c))}{4\pi n_0 e/c} \\[2ex] \dfrac{eB_0}{c}\dfrac{((i\omega/c) - (ik^2 c/\omega))}{4\pi n_0 e/c} & \dfrac{i\omega m_e((i\omega/c) - (ik^2 c/c))}{4\pi n_0 e/c} + e \end{pmatrix}$$

$$\times \begin{pmatrix} E_x \\ E_y \end{pmatrix} = \begin{pmatrix} 0 \\ 0 \end{pmatrix}.$$

Setting the determinant to 0 and letting $\Omega_e = -eB_0/mc$

$$\left(1 + \frac{k^2 c^2}{\omega_e^2} - \frac{\omega^2}{\omega_e^2}\right)^2 = \frac{\Omega_e^2}{\omega_e^4}\left(\omega - \frac{k^2 c^2}{\omega^2}\right)^2$$

$$\Rightarrow 1 = \left(\frac{\omega}{\omega_e^2} \pm \frac{\Omega_e}{\omega_e^2}\right)\left(\omega - \frac{k^2 c^2}{\omega^2}\right)$$

$$\Rightarrow \eta^2 \equiv \frac{k^2 c^2}{\omega^2} = 1 - \frac{\omega_e^2/\omega^2}{1 \pm \Omega_e/\omega}.$$

We can now define the R wave and the L wave.

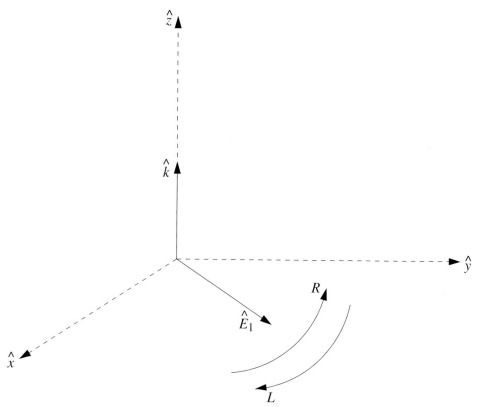

Fig. 10.2 Definition of Faraday rotation. The electric field vector rotates in the x–y plane. The rotation can be in either the right-hand or the left-hand sense.

R wave:

$$\eta^2 = \frac{c^2}{V_R^2} = 1 - \frac{w_e^2/w^2}{1 + \Omega_e/w}.$$
(267)

Right circularly polarized.

L wave:

$$\eta^2 = \frac{c^2}{V_L^2} = 1 - \frac{w_e^2/w^2}{1 - \Omega_e/w}.$$
(268)

Defining the phase speeds, we see that

$$\left(\frac{w}{k}\right)_R > \left(\frac{w}{k}\right)_L.$$

The implication of this inequality is that the plane of polarization rotates (Fig. 10.2).

$$\Rightarrow \text{Faraday Rotation.}$$

10.2.1 Rotation measure

Connecting theory to astronomical observations we note that the amount of rotation of the polarization vector depends on the difference in phase velocities of the R wave and the L wave. We can therefore proceed as we did for the dispersion measure.

$$\Delta t = t_R - t_L = \frac{l}{V_R} - \frac{l}{V_L} \tag{269}$$

where

$$\frac{c}{V_R} = \sqrt{1 - \frac{w_e^2/w^2}{1 + \Omega_e/w}}$$

$$\frac{c}{V_L} = \sqrt{1 - \frac{w_e^2/w^2}{1 - \Omega_e/w}}$$

from (267) and (268). In the limit of $|\Omega_e| \ll \omega$

$$\frac{c}{V} \approx 1 - \frac{1}{2}\frac{w_e^2/w^2}{1 \pm \Omega_e/w}. \tag{270}$$

Inserting into (269) yields

$$\Delta t \approx \frac{l}{2c}\frac{\omega_e^2}{\omega^2}\left[\frac{1}{1 - \Omega_e/w} - \frac{1}{1 + \Omega_e/w}\right]. \tag{271}$$

Cross multiplying and keeping the lowest order term in the denominator yields

$$\Delta t \approx \frac{\omega_e^2 \Omega_e l}{c\omega^3}$$

$$= \frac{2\pi e^3}{\omega^3 m_e^2 c^2} n_e B_0 l.$$

Now, the amount of rotation of E vector, in radians, is

$$\Psi = \omega \Delta t = \frac{2\pi e^3}{\omega^2 m_e^2 c^2} n_e B l.$$

Generalizing the quantity $n_e B_0 l$

$$\Psi = \frac{2\pi e^3}{\omega^2 m_e^2 c^2}\int_0^l n_e B_\parallel \, dl' = \frac{2.36 \times 10^4}{v^2}\int_0^l n_e B_\parallel \, dl'.$$

In terms of the wavelength we can define

$$\frac{\Psi}{\lambda^2} = \text{RM}$$

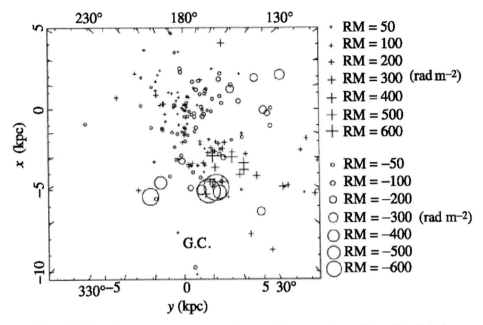

Fig. 10.3 Rotation measures measured toward known pulsars (from Rand, J. R. and Lyne, A. G., 1994, MNRAS, 268, pp. 497–505) and plotted such that they are viewed from the Galactic pole.

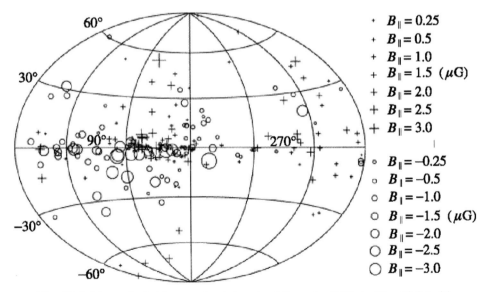

Fig. 10.4 Magnetic field strengths determined from the RMs in Fig. 10.3 (with the help of DMs). The projection is on the Galactic coordinate system as projected onto the sky (from Rand, J. R. and Lyne, A. G., 1994, MNRAS, 268, pp. 497–505).

where RM is the *rotation measure* and is defined as

$$\text{RM} = 2.62 \times 10^{-13} \int_0^l n_e B_{\parallel} \, dl' \text{ rad m}^{-2}. \tag{272}$$

Rotation measures can be determined empirically from polarization measurements made at different wavelengths, in the direction of pulsars (Fig. 10.3). They can be used, in conjunction with the dispersion measure, to determine the magnetic field along the line of sight to a pulsar, or for that matter any other radio source. Since only the line-of-sight component is relevant the above B is usually denoted as $B_{\parallel}$. An example of how the Galactic magnetic field is measured, using rotation measure, is given in Fig. 10.4.

10.3   Reference

Rand, R. J. and Lyne, A. G. (1994) New Rotation Measures of Distant Pulsars in the Inner Galaxy and Magnetic Field Reversals, *Monthly Notices of the Royal Astronomical Society*, **268**, pp. 497–505.

10.4   Further reading

Chen, F. (1974) *Introduction to Plasma Physics*, Plenum, New York, NY, USA.
Jackson, J. D. (1998) *Classical Electrodynamics*, John Wiley and Sons, New York, NY, USA.
Nicholson, D. R. (1983) *Introduction to Plasma Theory*, John Wiley and Sons, New York, NY, USA.
Shu, F. H. (1992) *The Physics of Astrophysics: Radiation*, University Science Books, Mill Valley, CA, USA.
Tucker, W. H. (1975) *Radiation Processes in Astrophysics*, MIT Press, Cambridge, MA, USA.
Verschuur, G. A. and Kellerman, K. I. (1988) *Galactic and Extragalactic Radio Astronomy*, Springer-Verlag, Berlin, Germany.

Part IV

Quantum mechanics

Astrophysical spectral lines offer two important insights into the workings of our Universe. First, they are probes of the fundamental (QM) nature of matter because they originate from subatomic, atomic and molecular systems. Second, they provide, via the Doppler effect, critical dynamical information on astrophysical systems ranging in scale from planetary systems to superclusters of galaxies. Examples of major contemporary problems in astrophysics that can be addressed through spectral line studies and the associated quantum mechanics include.

Missing mass and the halos of galaxies The most common element in the Universe is hydrogen and much of it is in a cold state. Given the 10 eV gap between the ground state and the first excited state of the simple Bohr atom, we should have little direct knowledge of this gas, yet it is the best studied gaseous component of the Universe. The reason is the 21 cm line corresponding to the hyperfine splitting of the ground state. The extremely low transition probability of this transition and the consequently narrow width of this line have led to its widespread use in measuring galaxy dynamics and kinematics. Studies of galaxy rotation have shown evidence for missing matter and point to the possibility of dark-matter halos. The nature of the dark matter and the implication on the long-term fate of the Universe remain contentious issues in astrophysics. The nature of this line and its use in these studies is discussed.

What is the ultimate fate of the stars? Analysis of debris left behind by dying stars provides clues to the fate of our own Sun. Classically forbidden spectral lines play a special role as diagnostics of these post-mortem plasmas which include planetary nebulas and supernova remnants.

How are stars born? Stars are believed to originate in dense molecular clouds. Molecular spectral lines are therefore instrumental in probing these stellar nurseries. Examples include molecules such as CO, OH and H_2.

What are quasars?   Among Nature's most luminous beacons quasars can be seen to the edges of the Universe but their true nature remains a mystery. I will explore the latest studies on these objects including the mystery of the "Lyman Forest".

Life in the Universe?   Spectral lines are relevant because of the following two questions: (1) What are the best frequencies to "listen" in on? (2) What evidence do we have that the molecular chains, needed for life to form on Earth, exist elsewhere?

Chapter 11

The hydrogen atom

A good starting point for a quantum mechanical understanding of spectral lines is the hydrogen atom. There are two good reasons for this approach. First, hydrogen is the most abundant element in the Universe, making up the bulk of the Universe both by mass and by volume. Any study of hydrogen is therefore relevant to much of the Universe. Second, hydrogen is the simplest of all atoms, consisting of one proton and one electron. Quantum theory assumes its simplest form and greatest accessibility for this simple element. The combination of quantum simplicity and the ubiquity of hydrogen makes the study of hydrogen a natural starting point for this part of the book.

11.1   Structure of the hydrogen atom

Since the H atom consists of only a proton and an electron, the Coulomb field experienced by the electron is spherically symmetric. Consequently, it is appropriate to solve Schrödinger's equation in spherical coordinates.

The time-independent Schrödinger equation is given by

$$H\phi = E\phi$$

$$H = -\frac{\hbar^2}{2m}\nabla^2 + V(r) \tag{273}$$

where

$$\nabla^2 = \left(\frac{\partial^2}{\partial r^2} + \frac{2}{r}\frac{\partial}{\partial r}\right) + \frac{1}{r^2}\left[\frac{1}{\sin\theta}\frac{\partial}{\partial\theta}\left(\sin\theta\frac{\partial}{\partial\theta}\right) + \frac{1}{\sin^2\theta}\frac{\partial^2}{\partial\phi^2}\right]$$

$$= \mathcal{R} + \frac{1}{r^2}\mathcal{L}^2. \tag{274}$$

Since $V(r)$ depends only on r, we anticipate a solution with separated variables. Thus, we can try something like

$$\phi(r, \theta, \phi) = R(r)Y(\theta, \phi). \tag{275}$$

Substituting (274) and (275) into (273)

$$\Rightarrow \left[-\frac{\hbar^2}{2m} \mathcal{R}R(r) \right] Y(\theta, \phi) - \frac{\hbar^2}{2mr^2} [\mathcal{L}^2 Y(\theta, \phi)] R(r)$$

$$+ V(r)R(r)Y(\theta, \phi) = ER(r)Y(\theta, \phi). \tag{276}$$

Multiplying each term by $((2mr^2)/\hbar^2)(1/(R(r)Y(\theta, \phi)))$

$$\Rightarrow \frac{r^2 \mathcal{R}R(r)}{R(r)} + \frac{2mr^2}{\hbar^2} [E - V(r)] = -\frac{\mathcal{L}^2 Y(\theta, \phi)}{Y(\theta, \phi)} = \Lambda$$

$$\Rightarrow \mathcal{R}R(r) + \frac{2m}{\hbar^2} [E - V(r)] R(r) = \frac{\Lambda}{r^2} R(r) \tag{277}$$

$$\mathcal{L}^2 Y(\theta, \phi) = -\Lambda Y(\theta, \phi) \tag{278}$$

where Λ = separation constant.

The function $R(r)$ carries important information on the electron's energy while $Y(\theta, \phi)$ carries information about the angular momentum.

The solution of (278) yields spherical harmonics and can be found in standard texts (for example Jackson, 1998) and we simply quote the result here.

$$Y(\theta, \phi) = Y_{\ell m}(\theta, \phi)$$

$$\Lambda = \ell(\ell + 1) \tag{279}$$

where ℓ is a positive integer, which allows us to reformulate the solution for ϕ as

$$\phi_{\ell m} = R(r)Y_{\ell m}(\theta, \phi). \tag{280}$$

Now, $V(r) = -e^2/r$ so that (277) becomes

$$\left(\frac{d^2}{dr^2} + \frac{2}{r}\frac{d}{dr} \right) R(r) - \frac{2m}{\hbar^2} \left[\left(\frac{\ell(\ell+1)\hbar^2}{2mr^2} - \frac{e^2}{r} \right) - E \right] R(r) = 0. \tag{281}$$

Solution of (281) yields the radial part of the wave function and the energy eigenstates. The latter are crucial for constructing the energy level diagram of hydrogen. Making the substitution $R(r) = g(r)/r$

$$\Rightarrow \frac{dR}{dr}(r) = \frac{dg(r)}{dr} - \frac{g(r)}{r^2}$$

$$\frac{d^2 R(r)}{dr^2} = \frac{d^2 g(r)}{dr^2} - \frac{2}{r^2}\frac{dg(r)}{dr} + \frac{2g(r)}{r^3}.$$

Substitute into (281)

$$\frac{\partial^2 g(r)}{\partial r^2} - \frac{2m}{\hbar^2}\left[\frac{\ell(\ell+1)\hbar^2}{2mr^2} - \frac{e^2}{r} - E\right]g(r) = 0. \tag{282}$$

Equation (282) cannot be solved in closed form.

Let us try some tricks. Let us construct solutions for the $r \to 0$ and $r \to \infty$ asymptotic limits then bridge the gap with a polynomial solution.

11.1.1 Case 1 $r \to \infty$

Equation (282) simplifies to

$$\frac{\partial^2 g(r)}{\partial r^2} + \frac{2m}{\hbar^2}Eg(r) = 0$$

yielding the solution

$$g_1(r) \propto e^{\pm(i/\hbar)\sqrt{2mE}r}. \tag{283}$$

Since we are looking for bound solutions ($E < 0$)

$$\Rightarrow \pm\frac{i}{\hbar}\sqrt{-2m\,|E|} = \mp\frac{1}{\hbar}\sqrt{2m\,|E|} = \pm\lambda$$

so that (283) becomes

$$g_1(r) \propto e^{\pm\lambda r} \propto e^{-\lambda r}[g(\infty) = 0].$$

11.1.2 Case 2 $r \to 0$

Equation (282) becomes

$$\frac{\partial^2 g(r)}{\partial r^2} - \frac{\ell(\ell+1)}{r^2}g(r) = 0.$$

Because of the $1/r^2$ term it is tempting to try

$$g_2(r) \propto r^p$$
$$\Rightarrow p(p-1)r^{p-2} - \ell(\ell+1)r^{p-2} = 0$$

which is only satisfied when

$$p = -\ell \qquad p = \ell + 1$$

$$\Rightarrow g_2(r) \propto \begin{cases} r^{-\ell}, & \ell \neq 0 \\ r^{\ell+1}. \end{cases}$$

But we want $g(0) = 0$

$$\Rightarrow g_2(r) \propto r^{\ell+1}. \tag{284}$$

11.1.3 What about the in-between?

Change to a dimensionless variable

$$\rho = 2\lambda r$$

$$\Rightarrow \frac{\partial^2 g(r)}{\partial r^2} = 4\lambda^2 \frac{\partial^2 g(\rho)}{\partial \rho^2}.$$

Substitute into (282)

$$\Rightarrow \frac{\partial^2 g(\rho)}{\partial \rho^2} - \left[\frac{\ell(\ell+1)}{\rho^2} - \frac{n}{\rho} + \frac{1}{4} \right] g(\rho) = 0 \tag{285}$$

where

$$n = \frac{me^2}{\hbar^2 \lambda} = \frac{1}{a_0 \lambda} \qquad \left[a_0 = \frac{\hbar^2}{m_e^2} \right]$$

where a_0 is the Bohr radius. Let us now try a general solution for $g(r)$ that is a product of $g_1(r)$, $g_2(r)$ and a power series to cover the in-between. With (283) and (284) we get

$$g(\rho) = e^{-\rho/2} \rho^{\ell+1} \sum_0^\infty a_j \rho^j.$$

Substitute into (285) to get the recursion relation

$$a_{k+1} = a_k \frac{\ell + k + 1 - n}{(k+1)(2\ell + k + 2)}.$$

A satisfactory solution must be normalized so that the infinite power series must be truncated to a polynomial. Truncation at the kth term $\rightarrow a_{k+1} = 0$. Thus, according to the previous recursion relation

$$a_{k+1} = 0 \Rightarrow n = \ell + k + 1.$$

Since k, ℓ are integers, n must also be an integer. It defines the principal quantum number of the atom and is restricted to the values

$$n = \ell + 1, \ell + 2, \ell + 3, \dots$$

which constrains ℓ to the values

$$\ell = 0, 1, 2, 3, \dots, n - 1.$$

The polynomial solution we are looking for takes the form of associated Laguerre polynomials

$$L_{n+\ell}^{2\ell+1}(\rho) = \left(\frac{d}{d\rho}\right)^{2\ell+1}\left[e^{\rho}\left(\frac{\partial}{\partial\rho}\right)^{n+\ell}(\rho^{n+\ell}e^{-\rho})\right]$$

so that

$$g(\rho) \propto e^{-\rho/2}\rho^{\ell+1}L_{n+\ell}^{2\ell+1}(\rho)$$

$$\Rightarrow R \propto \frac{g}{\rho} \propto e^{-\rho/2}\rho^{\ell}L_{n+\ell}^{2\ell+1}(\rho)$$

where

$$\rho = 2\lambda r = \frac{2}{na_0}r.$$

11.1.4 Normalizing R(r)

We note the integral

$$\int_0^\infty e^{-\rho}\rho^{2\ell}\left[L_{n+\ell}^{2\ell+1}(\rho)\right]^2 \rho^2 \, d\rho$$

$$= \frac{2n\left[(n+\ell)!\right]^3}{(n-\ell-1)!} \tag{286}$$

then

$$\int_0^\infty [R_{n\ell}(r)]^2 \, r^2 \, dr = \left(\frac{1}{2\lambda}\right)^3 \int_0^\infty [R_{n\ell}(\rho)]^2 \, \rho^2 \, d\rho$$

$$= \left(\frac{na_0}{2}\right)^3 \int_0^\infty [R_{n\ell}(\rho)]^2 \, \rho^2 \, d\rho. \tag{287}$$

Combining the expression for $g(\rho)$ with (286) and (287)

$$R_{n\ell}(r) = -2\left(\frac{1}{na_0}\right)^{3/2}\sqrt{\frac{(n-\ell-1)!}{n\left[(n+\ell)!\right]^3}}\left(\frac{2r}{na_0}\right)^{\ell}e^{-r/na_0}L_{n+\ell}^{2\ell+1}\left(\frac{2r}{na_0}\right). \tag{288}$$

These are the radial components of the electron wave functions. Table 11.1 lists the radial components up to $n = 3$. Figure 11.1 shows the geometry used to define the spatial variables, r, θ, ϕ. Figure 11.2 shows, graphically, all possible radial wave functions up to $n = 3$.

Table 11.1. *Wave functions and their components*

n	ℓ	m	$R_{n\ell}$	$Y_{\ell m}$	$|\phi_{n\ell m}|$
1	0	0	$2\left(\frac{1}{a_0}\right)^{3/2} e^{-r/a_0}$	$\frac{1}{2}\sqrt{\frac{1}{\pi}}$	$\frac{1}{\sqrt{\pi}}\left(\frac{1}{a_0}\right)^{3/2} e^{-r/a_0}$
2	0	0	$\left(\frac{1}{2a_0}\right)^{3/2}\left(2 - \frac{r}{a_0}\right) e^{-r/2a_0}$	$\frac{1}{2}\sqrt{\frac{1}{\pi}}$	$\frac{1}{4\sqrt{2\pi}}\left(\frac{1}{a_0}\right)^{3/2}\left(2 - \frac{r}{a_0}\right) e^{-r/2a_0}$
2	1	0	$\left(\frac{1}{2a_0}\right)^{3/2}\frac{1}{\sqrt{3}}\frac{r}{a_0} e^{-r/2a_0}$	$\frac{1}{2}\sqrt{\frac{3}{\pi}}\cos\theta$	$\frac{1}{4\sqrt{2\pi}}\left(\frac{1}{a_0}\right)^{3/2}\frac{r}{a_0} e^{-r/2a_0}\cos\theta$
2	1	±1	$\left(\frac{1}{2a_0}\right)^{3/2}\frac{1}{\sqrt{3}}\frac{r}{a_0} e^{-r/2a_0}$	$\mp\frac{1}{2}\sqrt{\frac{3}{2\pi}}\sin\theta\, e^{\pm i\phi}$	$\frac{1}{8}\sqrt{\frac{1}{\pi}}\left(\frac{1}{a_0}\right)^{3/2}\frac{r}{a_0} e^{-r/2a_0}\sin\theta\, e^{\pm i\phi}$
3	0	0	$2\left(\frac{1}{3a_0}\right)^{3/2}\left(1 - \frac{2}{3}\frac{r}{a_0} + \frac{2}{27}(r/a_0)^2\right) e^{-r/3a_0}$	$\frac{1}{2}\sqrt{\frac{1}{\pi}}$	$\frac{1}{81\sqrt{3\pi}}\left(\frac{1}{a_0}\right)^{3/2}\left(27 - 18\frac{r}{a_0} + 2(r/a_0)^2\right) e^{-r/3a_0}$
3	1	0	$\left(\frac{1}{3a_0}\right)^{3/2}\frac{4\sqrt{2}}{3}\left(1 - \frac{1}{6}\frac{r}{a_0}\right)\frac{r}{a_0} e^{-r/3a_0}$	$\frac{1}{2}\sqrt{\frac{3}{\pi}}\cos\theta$	$\frac{1}{81}\sqrt{\frac{2}{\pi}}\left(\frac{1}{a_0}\right)^{3/2}\left(6 - \frac{r}{a_0}\right)\frac{r}{a_0} e^{-r/3a_0}\cos\theta$
3	1	±1	$\left(\frac{1}{3a_0}\right)^{3/2}\frac{4\sqrt{2}}{3}\left(1 - \frac{1}{6}\frac{r}{a_0}\right)\frac{r}{a_0} e^{-r/3a_0}$	$\mp\frac{1}{2}\sqrt{\frac{3}{2\pi}}\sin\theta\, e^{\pm i\phi}$	$\frac{1}{8\sqrt{\pi}}\left(\frac{1}{a_0}\right)^{3/2}\left(6 - \frac{r}{a_0}\right)\frac{r}{a_0} e^{-r/3a_0}\sin\theta\, e^{\pm i\phi}$
3	2	0	$\left(\frac{1}{3a_0}\right)^{3/2}\frac{2\sqrt{2}}{27\sqrt{5}}\left(\frac{r}{a_0}\right)^2 e^{-r/3a_0}$	$\frac{1}{4}\sqrt{\frac{5}{\pi}}(3\cos^2\theta - 1)$	$\frac{1}{81\sqrt{6\pi}}\left(\frac{1}{a_0}\right)^{3/2}\frac{r^2}{a_0^2} e^{-r/3a_0}(3\cos^2\theta - 1)$
3	2	±1	$\left(\frac{1}{3a_0}\right)^{3/2}\frac{2\sqrt{2}}{27\sqrt{5}}\left(\frac{r}{a_0}\right)^2 e^{-r/3a_0}$	$\mp\frac{1}{2}\sqrt{\frac{15}{2\pi}}\sin\theta\cos\theta\, e^{\pm i\phi}$	$\frac{1}{81\sqrt{\pi}}\left(\frac{1}{a_0}\right)^{3/2}\left(\frac{r}{a_0}\right)^2 e^{-r/3a_0}\sin\theta\cos\theta\, e^{\pm i\phi}$
3	2	±2	$\left(\frac{1}{3a_0}\right)^{3/2}\frac{2\sqrt{2}}{27\sqrt{5}}\left(\frac{r}{a_0}\right)^2 e^{-r/3a_0}$	$\frac{1}{4}\sqrt{\frac{15}{2\pi}}\sin^2\theta\, e^{\pm 2i\phi}$	$\frac{1}{162\sqrt{\pi}}\left(\frac{1}{a_0}\right)^{3/2}\left(\frac{r}{a_0}\right)^2 e^{-r/3a_0}\sin^2\theta\, e^{\pm 2i\phi}$

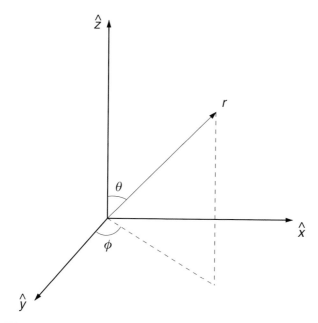

Fig. 11.1 The coordinate system used to define the variables in Table 11.1.

11.2   Total wave function

Substituting (288) into (275) and (279) yields the total solution

$$\phi_{n\ell m} = R_{n\ell}(r)Y_{\ell m}(\theta, \phi). \tag{289}$$

The simplest example corresponds to $n = 1, l = 0, m = 0$

$$\Rightarrow \phi_{100} = \frac{1}{\sqrt{\pi}} \left(\frac{1}{a_0}\right)^{3/2} e^{-r/a_0}.$$

The nomenclature associated with l is given by

$$\ell = 0, 1, 2, 3, 4 \ldots$$
$$\ell = s, p, d, f, g \ldots$$

The m's represent degenerate states. Table 11.1 lists the total wave functions up to $n = 3$, along with their radial and angular parts.

11.3   Probability functions

The probability of finding an electron in a unit volume dV is given by

$$P \, dV = \phi_{n\ell m}\phi_{n\ell m}^* \, dV.$$

"Images" of these probability functions are shown in Fig. 11.3.

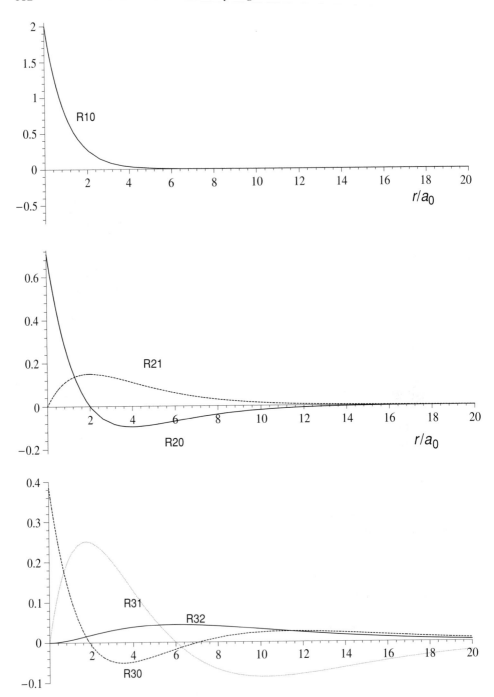

Fig. 11.2 Radial wave functions for the first three principal quantum numbers of hydrogen.

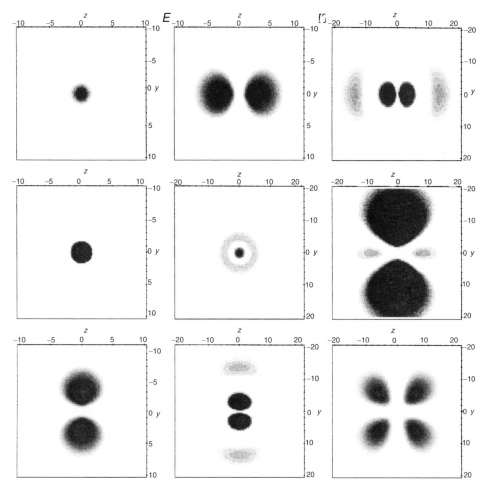

Fig. 11.3 Probability distributions (per unit volume). A slice through the $\phi = 0$ plane. The darker the shading the greater the probability density. Left column: from top to bottom, 1s, 2s, 2p0. Center column: 2p1, 3s, 3p0. Right column: 3p1, 3d0, 3d1.

In the case of spherical geometry, we can evaluate the probability of finding an electron in a shell of radius r, as follows

$$P \, \mathrm{d}V = |\phi_{n\ell m}|^2 \, \mathrm{d}V = |\phi_{n\ell m}|^2 \, r^2 \, \mathrm{d}r \, \mathrm{d}\Omega$$

$$\Rightarrow P(r) \, \mathrm{d}r = \int_{\Omega} |\phi_{n\ell m}|^2 \, r^2 \, \mathrm{d}r \, \mathrm{d}\Omega$$

$$= |R_{n\ell}(r)|^2 \, r^2 \, \mathrm{d}r \times \int_{\Omega} |Y_{\ell m}|^2 \, \mathrm{d}\Omega$$

$$P(r) \, \mathrm{d}r = r^2 \, |R_{n\ell}(r)|^2 \, \mathrm{d}r. \tag{290}$$

All possible radial probability functions, up to $n = 3$, are shown in Fig. 11.4.

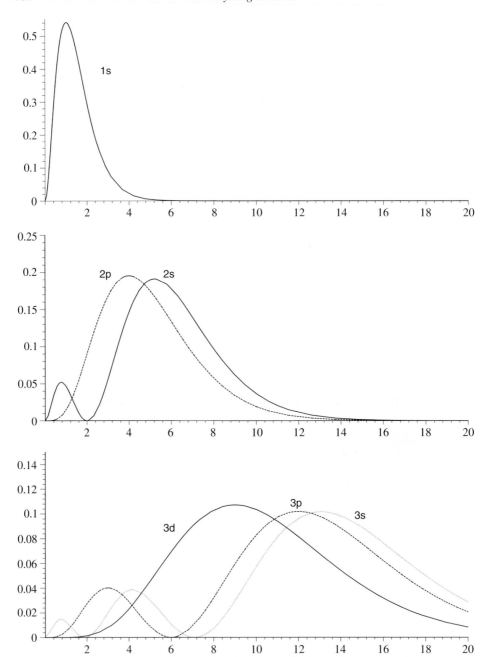

Fig. 11.4 The radial probability function $r^2 R^2$ which gives the relative probability of finding the electron at a given distance r/a_0.

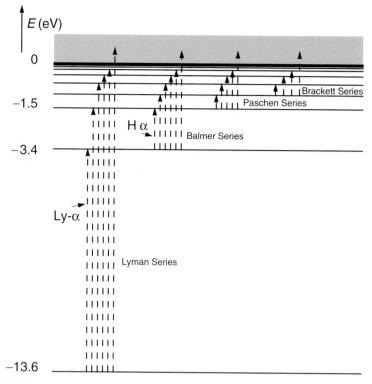

Fig. 11.5 The energy ladder for hydrogen based on its eigenstates. The well-known series of transitions are shown.

11.4 Energy eigenstates and transitions

Recalling the definition of λ

$$\lambda = \frac{1}{\hbar}\sqrt{2m\,|E|} = \frac{1}{na_0}$$

$$\Rightarrow E = E_n = \frac{\hbar^2}{m2n^2a_0^2}$$

$$\Rightarrow E_n = \frac{13.6}{n^2}\,\text{eV}. \tag{291}$$

These are the energy eigenstates of the hydrogen atom. These states are shown schematically in Fig. 11.5. Common transitions between these states are also shown.

11.5 Further reading

Anderson, E. E. (1971) *Modern Physics and Quantum Mechanics*, W. B. Saunders, Philadelphia, PA, USA.

Cohen-Tanoudji, C. (1992) *Principles of Quantum Mechanics*, John Wiley and Sons, New
   York, NY, USA.
Griffiths, D. (1995) *Introduction to Quantum Mechanics*, Prentice Hall, Englewood Cliffs,
   NJ, USA.
Hansch, T. S., Schawlow, A. L., Series, G. W. (1979) The Spectrum of the Hydrogen
   Atom, *Scientific American*, March, p. 94.
Landau, L. D. and Lifschitz, E. M. (1977) *Quantum Mechanics*, 3rd edn, Pergamon Press,
   Oxford, UK.
Landau, R. H. (1996) *Quantum Mechanics II: A Second Course in Quantum Theory*, 2nd
   edn, John Wiley and Sons, New York, NY, USA.
Osterbrock, D. E. (1974) *Astrophysics of Gaseous Nebulae*, W. H. Freeman, San
   Francisco, CA, USA.
Shu, F. H. (1992) *The Physics of Astrophysics: Radiation*, University Science Books, Mill
   Valley, CA, USA.
Tucker, W. H. (1975) *Radiation Processes in Astrophysics*, MIT Press, Cambridge, MA,
   USA.

Chapter 12

The interaction of radiation with matter

Now that we have discussed the structure of the hydrogen atom, we are able to examine the interaction of hydrogen atoms with photons from a quantum mechanical point of view. First, we begin with a non-relativistic treatment. Then, we will describe the single particle Hamiltonian with time varying EM fields. Perturbation analysis will then be used to separate the interaction part of the Hamiltonian from the static part.

12.1 Non-relativistic treatment

Recall from the last chapter that the H atom in its ground state has a radius given by the Bohr radius, a_0

$$a_0 = \frac{\hbar^2}{m_e e^2}$$

so that

$$\frac{m_e v^2}{2} = \frac{e^2}{2a_0} \rightarrow \frac{v}{c} = \frac{e^2}{\hbar c} = \alpha \approx 1/137$$

$$\Rightarrow v \ll c.$$

Thus, from the atom's point of view, the non-relativistic treatment is justified. Now let us examine the photon's point of view.

The highest photon energies that can interact with the H atom correspond to the ionization potential. Recall that

$$E_0 = \frac{e^2}{2a_0} = \hbar\omega$$

so that

$$k = \frac{\omega}{c} = \frac{\alpha}{2} \frac{1}{a_0}$$

$$\Rightarrow ka_0 = \frac{\alpha}{2} \ll 1$$

Since $ka_0 \ll 1$, multipole expansions converge rapidly so that the radiation field can be approximated by relatively few terms.

Now, let us find an expression for the single particle Hamiltonian that includes the time varying radiation field.

12.2 Single particle Hamiltonian

Borrowing from classical mechanics we can express the equation of motion of a particle in an EM field using the Lagrangian L so that

$$\frac{d}{dt}\left(\frac{\partial L}{\partial \dot{\vec{x}}}\right) = \frac{\partial L}{\partial \vec{x}}. \tag{292}$$

For a charged particle in a time varying EM field, the solution is given by

$$L = \frac{1}{2}m_e|\dot{\vec{x}}|^2 + e\Phi(\vec{x}, t) - \frac{e\dot{\vec{x}}}{c} \cdot \vec{A}(\vec{x}, t). \tag{293}$$

Substitution of this into (292) yields

$$m_e\ddot{\vec{x}} = -e\left(\vec{E} + \frac{\dot{\vec{x}}}{c} \times \vec{B}\right). \tag{294}$$

Now let us recall the definition of the canonical momentum

$$\vec{p} = \frac{\partial L}{\partial \dot{\vec{x}}} = m_e\dot{\vec{x}} - \frac{e}{c}\vec{A}. \tag{295}$$

To get the Hamiltonian, recall that

$$H \equiv \dot{\vec{x}} \cdot \vec{p} - L. \tag{296}$$

Taking (293) and (295) and substituting into (296) we get

$$H = m_e|\dot{\vec{x}}|^2 - \frac{e}{c}\dot{\vec{x}} \cdot \vec{A} - \frac{1}{2}m_e|\dot{\vec{x}}|^2 - e\Phi + \frac{e}{c}\dot{\vec{x}} \cdot \vec{A} = \frac{1}{2}m_e|\dot{\vec{x}}|^2 - e\Phi. \tag{297}$$

Using (295) again

$$H = \frac{1}{2m_e}\left|\vec{p} + \frac{e}{c}\vec{A}(\vec{x}, t)\right|^2 - e\Phi(\vec{x}, t). \tag{298}$$

12.3 Separation of static and radiation fields

This Hamiltonian is cumbersome to use. However, as I will now show, we can separate the static and radiation fields, which allows us to separate the Hamiltonian into a static part and radiative part. Consider an electron interacting with (a) an electrostatic field associated with an external charge distribution (usually the field of the nucleus and other electrons) and (b) vacuum radiation fields (photons).

Our goal is to choose a gauge in which the static and radiation fields can be separated. The Coulomb Gauge will do that as I will now demonstrate. The Coulomb Gauge is defined by the condition

$$\vec{\nabla} \cdot \vec{A} = 0. \tag{299}$$

Note that under such a condition $\vec{A}$ can be decomposed as

$$\vec{A} = \vec{\nabla} P + \vec{\nabla} \times \vec{Q} = \vec{A}_{\parallel} + \vec{A}_{\perp}.$$

We note the fact that we will usually deal with radiation fields that have transverse E and B fields, so that $\vec{A}_{\parallel} = 0$.

With this in mind, we can expand and interpret (298)

$$
\begin{aligned}
H &= \frac{1}{2m_e} \left[\vec{p} + \frac{e}{c}\vec{A}(\vec{x}, t) \right] \cdot \left[\vec{p} + \frac{e}{c}\vec{A}(\vec{x}, t) \right] - e\Phi(\vec{x}, t) \\
&= \frac{1}{2m_e} |\vec{p}|^2 - e\Phi(\vec{x}) + \frac{e}{2m_e c}(\vec{p} \cdot \vec{A} + \vec{A} \cdot \vec{p}) + \frac{e^2}{2m_e c^2} \vec{A} \cdot \vec{A} \tag{300} \\
&= H_0 + H_1 + H_2.
\end{aligned}
$$

12.3.1 Relative importance of H_0, H_1 and H_2

To demonstrate that we can use a perturbation analysis let us evaluate the ratios H_2/H_1 and H_1/H_0.

$$\frac{H_2}{H_1} \approx \frac{e^2 A^2/2m_e c^2}{epA/m_e c} = \frac{\alpha^2 a_0 A}{2ev/c} = \frac{\alpha a_0 A}{2e}. \tag{301}$$

For radiation we know that $\vec{B} = \nabla \times \vec{A} = i\vec{k} \times \vec{A}$ so that

$$A = \frac{B}{k} \approx \frac{2a_0 E}{\alpha}(B \approx E).$$

Also recall that $k \approx \alpha/(2a_0)$. Substituting into (301), we get

$$\frac{H_2}{H_1} \approx \frac{a_0^2 E}{e} = \frac{E_{rad}}{E_{static}}$$

$$\Rightarrow \frac{H_2}{H_1} \approx \left[2\pi n_{ph} a_0^3 \right]^{1/2} \tag{302}$$

where $(n_{ph} e^2)/2a_0 = n_{ph} \hbar\omega = E^2/4\pi$.

Now, the number of photons in a *Bohr cube* is roughly $n_{ph} a_0^3$ so when $n_{ph} a_0^3 \ll 1$ the ratio in equation (302) is small

$$\Rightarrow n_{ph} \ll a_0^{-3} = 10^{24} \text{ cm}^{-3}.$$

In stars, $n_{ph} \approx 10^{14}$ cm^{-3} so that the two-photon process is not that important.

As for the ratio H_0/H_1, note that

$$\frac{H_0}{H_1} \approx \frac{e}{a_0^2 E} \gg 1. \tag{303}$$

In summary

$$H_2 \ll H_1 \ll H_0 \tag{304}$$

We see that we are fully justified in using a perturbation analysis. With that in mind let us now derive an expression for H_1 so that we can quantify the transitions that result from the interaction of light with atoms.

12.4 Radiative transitions

12.4.1 Semi-classical approach

In this approach we treat the radiation field classically and the structure of the atom quantum mechanically. The latter has already been done, the former is justified by the conditions we have just derived.

The vector potential, $\vec{A}$, represents a time varying function associated with the EM field of the radiation field. Since photons have a wave-like nature (they also have a particle nature but in the above perturbation analysis we showed that $k \gg a_0$ so the wave description is sufficient for the following analysis) we can decompose $\vec{A}$ into its Fourier components so that

$$\vec{A}(\vec{x}, t) = \sum_k \left[\vec{a}(\vec{k}) \, e^{i(\vec{k}\cdot\vec{x} - \omega t)} + c.c \right]. \tag{305}$$

Given the Coulomb Gauge, $\nabla \cdot \vec{A} = 0$, we see from the above that $\vec{k} \cdot \vec{a}(\vec{k}) = 0$. We see that $\vec{a}(\vec{k})$ must have two components orthogonal to $\vec{k}$! These are, of course, the polarization (spin) states of the radiation field. Thus, a more complete description of the above is

$$\vec{A}(\vec{x}, t) = \sum_{k,\alpha} \left[\hat{e}_\alpha(\hat{k}) a_\alpha(\vec{k}) \, e^{i(\vec{k}\cdot\vec{x}-\omega t)} + c.c \right] \tag{306}$$

where $c.c$ indicates the complex conjugate of the preceding term and $\hat{e}_1$ and $\hat{e}_2$ are $\perp \vec{k}$.

12.4.2 The Hamiltonian of the radiation field

Recall that the Hamiltonian measures the total energy which, for a radiation field, is

$$H_{rad} = \frac{1}{8\pi} \int (|\vec{E}|^2 + |\vec{B}|^2) \, d^3x. \tag{307}$$

To evaluate the Hamiltonian of the radiation field we must borrow from Maxwell's equations the relations relating $\vec{E}$ and $\vec{B}$ to the vector potential, $\vec{A}$

$$\vec{E} = -\frac{1}{c}\frac{\partial \vec{A}}{\partial t} \quad \text{and} \quad \vec{B} = \vec{\nabla} \times \vec{A}.$$

Inserting (306) into this leads to the relations

$$\vec{E} = \sum_{k,\alpha} \left[\vec{E}_\alpha(\vec{k}) \, e^{i(\vec{k}\cdot\vec{x}-\omega t)} + c.c \right] \tag{308}$$

$$\vec{B} = \sum_{k,\alpha} \left[\vec{B}_\alpha(\vec{k}) \, e^{i(\vec{k}\cdot\vec{x}-\omega t)} + c.c \right] \tag{309}$$

where

$$\vec{E}_\alpha(\vec{k}) = -\frac{1}{c}(\hat{e}_\alpha a_\alpha(-i\omega)) = \frac{\omega}{c}(i\hat{e}_\alpha a_\alpha) = i k a_\alpha(\vec{k}) \hat{e}_\alpha$$
$$\vec{B}_\alpha(\vec{k}) = i k a_\alpha(\vec{k})(\hat{k} \times \hat{e}_\alpha).$$

Combining (307), (308) and (309) yields

$$H_{rad} = \frac{2V}{8\pi} \sum_{k,\alpha} |\vec{E}_\alpha(\vec{k})|^2 + |\vec{B}_\alpha(\vec{k})|^2 = \sum_{k,\alpha} |a_\alpha(\vec{k})|^2 \frac{k^2 V}{2\pi}. \tag{310}$$

If we now express H_{rad} in terms of the photon occupation number, $N_\alpha(\vec{k})$ (the spectrum of the radiation field), we expect

$$H_{rad} = \sum_{\vec{k},\alpha} \hbar\omega N_\alpha(\vec{k}). \tag{311}$$

Comparing (310) and (311) we see that

$$|a_\alpha(\vec{k})| = c \left[\frac{h N_\alpha(\vec{k})}{V\omega}\right]^{1/2}. \tag{312}$$

Equation (312) is relevant to the process of photon absorption by an atom. We can write down a similar expression for emission but we have to remember that emission can be spontaneous as well as stimulated. Without going through all the details, the emission process requires the transformation

$$N_\alpha(\vec{k}) \rightarrow 1 + N_\alpha(\vec{k}).$$

12.4.3 The perturbation Hamiltonian

We are now ready to define H_1. Combining (298), (306) and (312), and noting that $\vec{p} \cdot \vec{A} \propto \vec{\nabla} \cdot \vec{A} = 0$, we have

$$H_1 = \sum_{\vec{k},\alpha} \left[H_\alpha^{abs}(\vec{k})\, e^{-i\omega t} + H_\alpha^{em}(\vec{k})\, e^{i\omega t}\right] \tag{313}$$

where

$$H_\alpha^{abs}(\vec{k}) = \frac{e}{m_e}\left[\frac{h}{\omega V} N_\alpha(\vec{k})\right]^{1/2} e^{i\vec{k}\cdot\vec{x}} \hat{e}_\alpha(\hat{k}) \cdot \vec{p} \tag{314}$$

and

$$H_\alpha^{em}(\vec{k}) = \frac{e}{m_e}\left[\frac{h}{\omega V}(1 + N_\alpha(\vec{k}))\right]^{1/2} e^{-i\vec{k}\cdot\vec{x}} \hat{e}_\alpha(\hat{k}) \cdot \vec{p}. \tag{315}$$

Finally, in the continuum limit

$$H_1 = \sum_{\alpha=1}^{2} \frac{V}{(2\pi)^3} \int \left[H_\alpha^{abs}(\vec{k})\, e^{-i\omega t} + H_\alpha^{em}(\vec{k})\, e^{i\omega t}\right] d^3k. \tag{316}$$

We have an expression for H_1. We can now go ahead and construct the total Hamiltonian, $H = H_0 + H_1$ (neglecting the two-photon process) and solve the Schrödinger equation. However, given the fact that an interaction of an atom with a photon is necessarily a time-dependent process we must solve the time-dependent Schrödinger equation.

12.4.4 Time-dependent perturbation theory

The time evolution of the electron wave function is governed by Schrödinger's equation

$$H\Psi = i\hbar\frac{\partial\Psi}{\partial t} \tag{317}$$

where $H = H_0 + H_1$ and $H_0 = -(\hbar^2/(2m_e))\nabla^2 - e\Phi(\vec{x})$. Recall that H_0 has eigenfunctions of the form

$$\Psi_j(\vec{x}, t) = \phi_j(\vec{x})\,e^{-iE_jt/\hbar} \tag{318}$$

where the energy, E_j, of the jth *eigenstate* is an *eigenvalue* of the time-independent Schrödinger equation

$$H_0\phi_j = E_j\phi_j. \tag{319}$$

For H_0, the set of all ϕ_j's forms a basis for any possible wave function for stationary atomic systems. It should therefore be possible to construct any perturbed (time-dependent) Ψ from

$$\Psi(\vec{x}, t) = \sum_j c_j(t)\phi_j(\vec{x})\,e^{-iE_jt/\hbar} \tag{320}$$

where $c_j(t)$ are evaluated according to the constraint represented by (317). Normally, Ψ is chosen to be normalized so that

$$\int \Psi^*\Psi\,d^3x = 1. \tag{321}$$

In Dirac *bra-ket* notation this becomes

$$\langle\Psi|\Psi\rangle = 1. \tag{322}$$

The eigenfunctions form an *orthonormal* set such that

$$\langle\phi_l|\phi_j\rangle = \delta_{lj}. \tag{323}$$

With the bra-ket introduction out of the way let us go back to solving the coefficients, c_j. Substituting (320) into (317) we get

$$(H_0 + H_1)\sum_j c_j(t)\phi_j(\vec{x})\,e^{-iE_jt/\hbar} = i\hbar\sum_j \phi_j(\vec{x})\frac{\partial}{\partial t}c_j(t)\,e^{-(iE_jt)/\hbar}. \tag{324}$$

Expanding and using (319) we end up with

$$\sum_j H_1 c_j(t)\phi_j(\vec{x})\,e^{-iE_jt/\hbar} = i\hbar\sum_j \dot{c}_j(t)\phi_j(\vec{x})\,e^{-iE_jt/\hbar}. \tag{325}$$

Multiplying both sides by $\phi_f^* \, e^{iE_f t/\hbar}$ gets the final state into the equation. Integrating over V results in

$$\sum_j e^{it(E_f - E_j)/\hbar} c_j(t) \int \phi_f^* H_1 \phi_j \, \mathrm{d}^3 x = i\hbar \dot{c}_j(t) \, e^{it(E_f - E_j)/\hbar} \int_V \phi_f^* \phi_j \, \mathrm{d}^3 x$$

which simplifies to

$$\sum_j e^{i\omega_{fj} t} c_j(t) \langle \phi_f | H_1 | \phi_j \rangle = i\hbar \dot{c}_f(t) \tag{326}$$

where

$$\omega_{fj} \equiv (E_f - E_j)/\hbar$$

is +ve for $E_f > E_j$ (absorption) and −ve for $E_f < E_j$ (emission).

If we take the initial conditions as

$$\Psi(\vec{x}, 0) = \phi_i(\vec{x}) \rightarrow c_j(0) = \delta_{ji} \tag{327}$$

we can iterate by assuming that $c_j = \delta_{ji}$ for *all* t. Substituting this into the left-hand side of (326) leads to

$$\sum_j e^{i\omega_{fj} t} \delta_{ji} \langle \phi_f | H_1 | \phi_j \rangle = i\hbar \dot{c}_f(t)$$

$$\Rightarrow e^{i\omega_{fi} t} \langle \phi_f | H_1 | \phi_i \rangle = i\hbar \dot{c}_f(t).$$

Integrating,

$$c_f(t) = -\frac{i}{\hbar} \int_0^t \langle \phi_f | H_1 | \phi_i \rangle \, e^{i\omega_{fi} t} \, \mathrm{d}t. \tag{328}$$

Substitute the new value into (326) and continue iterating . . .

12.5   Absorption of photons

To see how the absorption of photons takes place, we will now derive an expression for the transition probability for absorption. We begin by considering one component of the absorption part of H_1,

$$H_\alpha^{abs}(\vec{k}) \, e^{-i\omega t}. \tag{329}$$

Replace H_1 with (329) and substitute into (328)

$$\Rightarrow c_f(\vec{k}, t) = -\frac{i}{\hbar} \int_0^t \langle \phi_f | H_\alpha^{abs} | \phi_i \rangle \, e^{i(\omega_{fi} - \omega)t} = -\frac{1}{\hbar} \langle \phi_f | H_\alpha^{abs} | \phi_i \rangle \frac{e^{i(\omega_{fi} - \omega)t} - 1}{\omega_{fi} - \omega}. \tag{330}$$

The absorption probability follows by squaring the above

$$|c_f(\vec{k}, t)|^2 = \hbar^{-2}|\langle \phi_f|H_\alpha^{abs}|\phi_i\rangle|^2 \frac{\sin^2[(\omega - \omega_{fi})t/2]}{[(w - \omega_{fi})/2]^2} \tag{331}$$

which is the probability of absorbing a photon of frequency $\omega = ck$!

To get the total transition probability we must sum over *all* k and α

$$P_{if} = \sum_{\vec{k}, \alpha} |c_f(\vec{k}, t)|^2$$

$$= \frac{V}{(2\pi)^3} \hbar^{-2} \sum_{\alpha=1}^{2} \int |\langle \phi_f|H_\alpha^{abs}(\vec{k})|\phi_i\rangle|^2 \frac{\sin^2[(\omega - \omega_{fi})t/2]}{[(\omega - \omega_{fi})/2]^2} \, d^3k. \tag{332}$$

The above can be cast into a more revealing form by substituting the absorption part of (316) and the relation $d^3k = k^2 \, dk \, d\Omega = \omega^2/c^3 \, d\omega \, d\Omega$ into

$$\Rightarrow P_{if} = \left(\frac{e}{2\pi m_e}\right)^2 \sum_{\alpha=1}^{2} \int \frac{N_\alpha(\vec{k})}{\hbar\omega} |\langle \phi_f| \, e^{i\vec{k}\cdot\vec{x}} \hat{e}_\alpha(\vec{k}) \cdot \vec{p}|\phi_i\rangle|^2$$

$$\times \frac{\sin^2[(\omega - \omega_{fi})t/2]}{[(\omega - \omega_{fi})/2]^2} \frac{\omega^2}{c^3} \, d\omega \, d\Omega. \tag{333}$$

Notice how the dependence on V dropped out. Note that for $t \gg 2/\omega_{fi}$ this function is strongly peaked. Thus

$$\Rightarrow \int_0^\infty F(\omega) \frac{\sin^2[(\omega - \omega_{fi})t/2]}{[(\omega - \omega_{fi})/2]^2} \, d\omega$$

$$\approx 2t \, F(\omega_{fi}) \int_{-\infty}^\infty \frac{\sin^2 \zeta}{\zeta^2} \, d\zeta = 2\pi t \, F(\omega_{fi}) \tag{334}$$

where $F(\omega)$ is any continuous function.

Applying (334) to (333) we get

$$P_{if} = t \left(\frac{e^2}{hc^3 m_e^2}\right) \sum_{\alpha=1}^{2} \int_\Omega [\omega N_\alpha(\vec{k}) \langle \phi_f| e^{i\vec{k}\cdot\vec{x}} \hat{e}_\alpha(\vec{k}) \cdot \vec{p}|\phi_i\rangle^2]_{fi} \, d\Omega. \tag{335}$$

This predicts that the transition will eventually occur. What we really want to know is what the transition probability rate is, so

$$\frac{\partial P_{if}}{\partial t} = \sum_{\alpha=1}^{2} \int \omega_\alpha \, d\Omega \tag{336}$$

where ω_α = a constant probability rate per $d\Omega$ for the radiative transition, $i \to f$.

So there we have it, a basis for calculating the transition rate between any two levels of an atom. It boils down to evaluating the matrix element $\langle \phi_f | e^{i\vec{k}\cdot\vec{x}} \hat{e}_\alpha(\vec{k}) \cdot \vec{p} | \phi_i \rangle$. In general, though, this is tough to do. However, if we expand $e^{i\vec{k}\cdot\vec{x}} \hat{e}_\alpha(\vec{k}) \cdot \vec{p}$ in a multipole expansion, the problem is greatly simplified. We do that next, beginning with the dipole approximation.

12.5.1 Absorption cross-sections

Now that we have an expression for the interaction Hamiltonian, H_1, let us formalize the expression for the absorption part of it by deriving expressions for the absorption cross-section. The cross-section is what is normally quoted in the literature so this is an attempt to connect our quantum mechanical development to tabulated properties of atoms. In the process we will discuss bound–bound and bound–free (ionization) absorption processes. The role of spontaneous emission and the natural line width will also be discussed.

The essential step in deriving an expression for the absorption cross-section is to evaluate the H_α^{abs} operator

$$\langle \phi_f | H_\alpha^{abs} | \phi_i \rangle = \frac{e}{m_c} \left[\frac{h}{V\omega} N_\alpha(\vec{k}) \right]^{1/2} \times \langle \phi_f | e^{i\vec{k}\cdot\vec{x}} \hat{e}_\alpha(\vec{k}) \cdot \vec{p} | \phi_i \rangle. \qquad (337)$$

As noted earlier this expression is difficult to solve without making an approximation. The logical thing to do, in light of the fact that we are in the non-relativistic regime, is to expand this complex exponential in a multipole expansion, so that

$$e^{i\vec{k}\cdot\vec{x}} = 1 + i\vec{k}\cdot\vec{x} + \cdots$$

In the dipole approximation

$$e^{i\vec{k}\cdot\vec{x}} \approx 1$$

because $\vec{k}\cdot\vec{x} \ll 1$.

Substituting into this and rearranging terms we have the following matrix element to solve

$$\hat{e}_\alpha \cdot \langle \phi_f | \vec{p} | \phi_i \rangle$$

which is much simpler than that we started with.

To simplify the calculation of the matrix element further we express $\vec{p}$ in terms of H_0, $\vec{x}$.

Use $[H_0, \vec{x}] = H_0\vec{x} - \vec{x}H_0 = -\hbar^2/2m_e[\nabla^2\vec{x} - \vec{x}\nabla^2] = -\hbar^2/m_e\vec{\nabla} = -i\hbar/m_e\vec{p}$
so that $\vec{p} = (im_e/\hbar)(H_0\vec{x} - \vec{x}H_0)$.

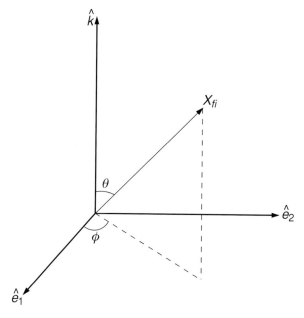

Fig. 12.1 The coordinate system associated with $\vec{X}$.

Thus

$$\langle \phi_f | \vec{p} | \phi_i \rangle = i m_e \omega_{fi} \vec{X}_{fi}$$

where

$$\vec{X}_{fi} \equiv \langle \phi_f | \vec{x} | \phi_i \rangle \tag{338}$$

and

$$-e\vec{X}_{fi} = \langle \phi_f | \vec{d} | \phi_i \rangle.$$

All we have to do now is to insert the above into (335) and (336) to get the transition probability rate.

12.5.2 Dipole transition probability

Substituting into (336) yields

$$\frac{\mathrm{d}P_{if}}{\mathrm{d}t} = \frac{e^2}{hc^3} \sum_{\alpha=1}^{2} \oint [N_a(\vec{k})\omega^3 |\hat{e}_\alpha(\vec{k}) \cdot \vec{X}_{fi}|^2]_{fi} \, \mathrm{d}\Omega. \tag{339}$$

We now have to consider the geometry of the interaction (Fig. 12.1). The vector $\vec{X}_{fi}$ represents the orientation of the atomic charge distribution while $\vec{k}$, $\hat{e}_1$ and $\hat{e}_2$

describe the orientation of the photon field and propagation vectors. Projecting $\vec{X}_{fi}$ into the $(\hat{e}_1, \hat{e}_2)$ plane yields

$$\hat{e}_1 \cdot \vec{X}_{fi} = |\vec{X}_{fi}| \sin\theta \cos\phi$$
$$\hat{e}_2 \cdot \vec{X}_{fi} = |\vec{X}_{fi}| \sin\theta \sin\phi.$$

For an isotropic, unpolarized field $N_1 = N_2 \equiv N(\omega)/2$.

If we now substitute this information into (339) and integrate over the angular variables we end up with

$$\frac{dP_{if}}{dt} = \frac{4\pi e^2 \omega_{fi}^3}{3hc^3} N(\omega_{fi})|\vec{X}_{fi}|^2. \tag{340}$$

Now we are in a position to define the bound–bound absorption cross-section, $\sigma_{bb}(\omega)$, by comparing (340) with the standard definition of the cross-section.

12.5.3   Bound–bound absorption cross-section

It is conventional to define $\sigma_{bb}(\omega)$ such that

$$\frac{dP_{if}}{dt} = \int_0^\infty \sigma_{bb}(\omega) \, cN(\omega) \frac{4\pi\omega^2 \, d\omega}{(2\pi)^3 c^3}. \tag{341}$$

Comparison with (340) yields

$$\sigma_{bb}(\omega) = \frac{4\pi^2}{3} \left(\frac{e^2}{\hbar c}\right) |\vec{X}_{fi}|^2 \omega \, \delta(\omega - \omega_{fi}). \tag{342}$$

The expression for the classically derived absorption coefficient is

$$\alpha_\nu = \frac{\pi e^2}{m_e c} f_{12} \phi_{12}(\nu). \tag{343}$$

If we now equate (342) and (343) we can identify the following relationships

$$\phi_{12}(\nu) = \omega\delta(\omega - \omega_{fi}) = 2\pi\nu\delta(\nu - \nu_{fi})$$
$$\Rightarrow f_{12} = (4\pi m_e/3\hbar)\nu_{21}|\vec{X}_{21}|^2 = 2m_e(\omega_{21}|\vec{X}_{21}|)^2/3\hbar\omega_{21}. \tag{344}$$

We now have a QM interpretation of oscillator strengths. This however is not quite complete because the expression for the line profile ϕ_{12} is suspect.

12.6 Spontaneous emission

What we failed to consider previously is that emission processes compete with the absorption processes so that the absorption rate is affected by the transition rate for spontaneous emission. Let us try and add that to get a more complete picture.

We begin by introducing the A coefficient as $A_{fi} = dP_{fi}/dt$. Let $N_\alpha \to 1$, reverse $if \to fi$ and proceed as before

$$A_{fi} = \frac{e^2}{hc^3} \sum_{\alpha=1}^{2} \oint [\omega^3 |\hat{e}_\alpha(\vec{k}) \cdot \vec{X}_{fi}|^2]_{fi} \, d\Omega. \tag{345}$$

Given that $|\vec{X}_{fi}| = |\vec{X}_{if}|$ and integrating

$$A_{fi} = \frac{4e^2\omega_{fi}^3}{3\hbar c^3} |\vec{X}_{fi}|^2.$$

Now let us go back to time-dependent perturbation theory and let us set the transition rate

$$\frac{d}{dt}(|c_f|^2)|_{spon} = -A_{fi}|c_f|^2 = -\Gamma|c_f|^2$$

$$\Rightarrow \dot{c}_f = -\frac{\Gamma c_f}{2}.$$

We can use this result to modify $\dot{c}_f(t)$ (absorption)

$$\dot{c}_f(t) \Rightarrow \underbrace{-\frac{i}{\hbar}\langle \phi_f | H_\alpha^{abs} | \phi_i \rangle e^{i(\omega_{fi} - \omega)t}}_{\text{absorption}} \underbrace{-\frac{\Gamma}{2}c_f}_{\text{emission}}. \tag{346}$$

Integrating

$$c_f(t) = -\frac{1}{\hbar}\langle \phi_f | H_\alpha^{abs} | \phi_i \rangle \left[\frac{e^{i(\omega_{fi} - \omega)t} - e^{-\Gamma t/2}}{\omega_{fi} - \omega - i\Gamma/2}\right]. \tag{347}$$

Multiplying by c.c

$$\Rightarrow |c_f(t)|^2 = \hbar^{-2} |\langle \phi_f | H_\alpha^{abs} | \phi_i \rangle|^2 \left[\frac{1 + e^{-\Gamma t} - 2e^{-\Gamma t/2}\cos(\omega_{fi} - \omega)t}{(\omega_{fi} - \omega)^2 + (\Gamma/2)^2}\right]. \tag{348}$$

In the limit $t \to \infty$ we have

$$|c_f(t)|^2 = \frac{\hbar^{-2}|\langle \phi_f | H_\alpha^{abs} | \phi_i \rangle|^2}{(\omega_{fi} - \omega)^2 + (\Gamma/2)^2}. \tag{349}$$

We can now define $\sigma_{bb}(\omega)$ as

$$\sigma_{bb}(\omega) = \frac{4\pi^2}{3}\left(\frac{e^2}{\hbar c}\right)|\vec{X}_{fi}|^2 \omega_{fi} \mathcal{L}(\omega)$$

where

$$\mathcal{L}(\omega) \equiv \frac{1}{\pi} \left[\frac{\Gamma/2}{(\omega - \omega_{fi})^2 + (\Gamma/2)^2} \right]$$

which is the Lorentzian profile associated with the natural line profile. The above represent the definition of the absorption cross-section which we have now defined in terms of the dipole transition element and the Lorentzian. I now consider a natural extension of this discussion, the bound–free transition. I first discuss the concept of photoionization and then we determine an analogous expression for the bound–free cross-section.

12.7   Photoionization

Let us now consider the bound–free process. Let us begin with

$$\frac{\mathrm{d}P_{if}}{\mathrm{d}t} = \frac{e^2}{hc^3 m_e^2} \sum_{\alpha=1}^{2} \oint [\omega N_\alpha(\vec{k}) |\langle \phi_f | e^{i\vec{k}\cdot\vec{x}} \hat{e}_\alpha \cdot \vec{p} | \phi_i \rangle|^2]_{fi} \, \mathrm{d}\Omega. \tag{350}$$

We now need to define the initial and final states. With an atom at the origin of the coordinate system, the initial state

$$\phi_i = \left(\pi a_z^3\right)^{-1/2} e^{-|\vec{x}|/a_z} \tag{351}$$

represents the bound state where

$$a_z = \frac{\hbar^2}{Z m_e e^2} \qquad \text{effective size of atom.}$$

To determine the final state, we use the Born approximation which makes the assumption that the electron is isolated in a vacuum once it is liberated from the atom, so that

$$\phi_f = V^{-1/2} e^{i\vec{k}_e \cdot \vec{x}} \Rightarrow \phi_f^* = V^{-1/2} e^{-i\vec{k}_e \cdot \vec{x}}. \tag{352}$$

Furthermore

$$\hbar\omega = \frac{p^2}{2m_e} = \left(\frac{\hbar}{\lambda}\right)^2 \frac{1}{2m_e} = E \Rightarrow |\vec{k}_e| = \frac{(2m E_f)^{1/2}}{\hbar} \tag{353}$$

where $E_f = \hbar\omega_{fi} - Ze^2/2a_z$.

Substituting (351), (352) and (353) into (350) and using $\vec{p} = -i\hbar\vec{\nabla}$, we get

$$\Rightarrow \frac{\hbar^2}{\pi a_z^3 V} \left| \int_V e^{i(\vec{k}-\vec{k}_e)\cdot\vec{x}} \hat{e}_\alpha \cdot \vec{\nabla}(e^{-|\vec{x}|/a_z}) \, \mathrm{d}^3 x \right|^2$$

the "square of the matrix element".

Integrating by parts

$$\Rightarrow \frac{\hbar^2}{\pi a_z^3 V} (\vec{k}_e \cdot \hat{e}_\alpha)^2 \left| \int_V e^{i(\vec{k}-\vec{k}_e)\cdot\vec{x} - |\vec{x}|/a_z} \, d^3x \right|^2$$

where $\vec{k} \cdot \hat{e}_\alpha = 0$ was used.

Let $\vec{q} = \vec{k} - \vec{k}_e \Rightarrow \hbar\vec{q} =$ momentum *not* transferred. Using spherical polar coordinates

$$\vec{q} \cdot \vec{x} = qr \cos\theta, \quad d^3x = 2\pi r^2 \sin\theta \, dr \, d\theta, \quad \mu = \cos\theta, \quad V \to \infty$$

$$\Rightarrow \int e^{i\vec{q}\cdot\vec{x} - |\vec{x}|/a_z} \, d^3x = 2\pi \int_0^\infty r^2 \, e^{-r/a_z} \int_{-1}^{+1} e^{iqr\mu} \, d\mu \, dr = 8\pi a_z^3 / (1 + q^2 a_z^2)^2.$$

Collecting all expressions

$$|\langle \phi_f | e^{i\vec{k}\cdot\vec{x}} \hat{e}_\alpha \cdot \vec{p} | \phi_i \rangle|^2 = 64\pi \frac{\hbar^2 a_z^3}{V} \frac{(\vec{k}_e \cdot \hat{e}_\alpha)^2}{(1 + q^2 a_z^2)^4}.$$

Substituting into (350) and dropping the fi subscripts on the right-hand side

$$\Rightarrow \frac{dP_{if}}{dt} = \frac{32e^2\hbar a_z^3}{Vc^3 m_e^2} \sum_{\alpha=1}^{2} \oint \frac{[\omega N_\alpha(\vec{k})(\hat{k}_e \cdot \hat{e}_\alpha)^2]}{[(1 + |\vec{k} - \vec{k}_e|^2 a_z^2)]^4} \, d\Omega. \qquad (354)$$

For an ensemble of randomly oriented atoms, photons appear isotropic with respect to e^- ejection $\vec{k}_e$ (Fig. 12.2) so that,

$$\vec{k}_e \cdot \hat{e}_\alpha = k_e \sin\theta \cos\phi$$
$$|\vec{k} - \vec{k}_e|^2 = k^2 + k_e^2 - 2kk_e \cos\theta.$$

Combine with (353) and $\omega = ck$

$$k_e^2 = \frac{2m_e ck}{\hbar} - \frac{1}{a_z^2} \qquad (355)$$

where

$$a_z = \frac{\hbar^2}{Zm_e e^2}.$$

Collecting expressions again

$$1 + |\vec{k} - \vec{k}_e|^2 a_z^2 = a_z^2 \frac{2m_e ck}{\hbar} \left(1 + \frac{\hbar k}{2m_e c} - \frac{\hbar k_e}{2m_e c} \cos\theta \right)$$

$$\approx a_z^2 \frac{2m_e \omega}{\hbar} \qquad \text{(non-relativistic)}.$$

Now we are in a position to evaluate the angular integral in dP_{if}/dt.

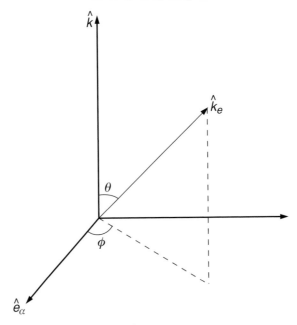

Fig. 12.2 The geometry that defines $\hat{k}_e$.

Thus

$$\int_0^{2\pi} d\phi \int_0^{\pi} \sin^2\theta \cos^2\phi \sin\theta \, d\theta = \frac{4\pi}{3} \tag{356}$$

$$\Rightarrow \frac{dP_{if}}{dt} = \frac{8\pi e^2 a_z^3}{3V\hbar m_e^2 c^3} \omega^{-3} N(\omega)(\hbar k_e)^2 \left(\frac{\hbar}{m_e a_z^2}\right)^4 \tag{357}$$

$$\qquad\qquad\qquad\qquad \uparrow \qquad \uparrow \qquad \uparrow$$
$$\qquad\qquad\qquad\quad \text{photons electron atom}$$

12.7.1 Bound–free cross-sections

Our goal is to determine a σ_{bf} analogous to σ_{bb}.

Let dN_f be the number of final free electron states between k_e and $k_e + dk_e$ so that

$$dN_f = \frac{V}{(2\pi)^3} 4\pi k_e^2 \, dk_e. \tag{357}$$

Define $\sigma_{bf}(\omega)$ via

$$\frac{dP_{if}}{dt} dN_f = \sigma_{bf}(\omega) \left[\frac{cN(\omega)4\pi\omega^2}{(2\pi)^3 c^3} \, d\omega\right]. \tag{358}$$

Let $k_e \, dk_e = (m_e/\hbar) \, d\omega$ and use (84) and (86)

$$\Rightarrow \sigma_{bf}(\omega) = \frac{8\pi}{3}\left(\frac{e^2}{m_e c}\right)\left(\frac{\hbar}{m_e a_z^2}\right)^4 \omega^{-5} (k_e a_z)^3 \tag{359}$$

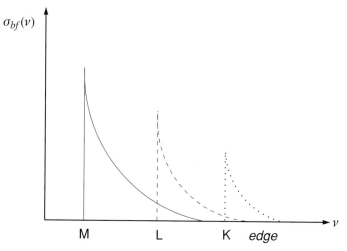

Fig. 12.3 Bound–free absorption cross-sections. These plots correspond to the M, L and K shell electrons. Note the abrupt rise at the ionization frequency followed by the $1/v^3$ fall-off.

where $k_e^2 = (2m_e/\hbar^2)(\hbar\omega - (Ze^2/2a_z)) = +\text{ive}$ if $\hbar\omega > IP$ and IP is the ioniza-tion potential. To use the Born approximation, let $\hbar\omega \gg Ze^2/2a_z$

$$\Rightarrow k_e \approx \left[\frac{2m_e\omega}{\hbar}\right]^{1/2}$$

which leads to

$$\sigma_{bf}(\omega) = \frac{8\pi}{3\sqrt{3}}\frac{Z^4 m_e e^{10}}{c\hbar^3(\hbar\omega)^3}\left[48\frac{Ze^2}{2a_z\hbar\omega}\right]^{1/2}. \tag{360}$$

The above can be compared to the standard formulation,

$$\alpha_v(n) = n^{-5}\frac{8\pi}{3\sqrt{3}}\frac{Z^4 m_e e^{10}}{c\hbar^3(hv)^3}g_{bf}(v).$$

Now we have a QM basis for b–b and b–f transitions. Plots of σ_{bf} are shown in Fig. 12.3.

12.8 Selection rules

Now that we have workable expressions for the interaction Hamiltonian and the structure Hamiltonian we are well placed to work out and understand the selection rules associated with electronic transitions. In the process we will discuss selection rules for electric dipole and quadrupole and magnetic dipole transitions, and the concept of forbidden transitions.

12.8.1 Dipole selection rules

Consider a radiative transition $n\ell m \rightarrow n'\ell'm'$. To describe it we need to evaluate

$$\langle \phi_f | \hat{e}_\alpha \cdot \vec{x} | \phi_i \rangle$$

which can be split into

$$\langle \phi_f | z | \phi_i \rangle$$
$$\frac{1}{\sqrt{2}} \langle \phi_f | x \pm iy | \phi_i \rangle$$

by considering the geometry of the interaction. In spherical coordinates

$$z = r \cos \theta \quad \text{and} \quad x \pm iy = r \sin \theta \, e^{\pm i\phi}.$$

Let

$$\mu = \cos \theta \quad \text{and} \quad d^3x = r^2 \, dr \, d\mu \, d\phi$$

$$\Rightarrow \langle \phi_f | z | \phi_i \rangle \propto \left[\int_0^\infty r^3 R_{n'\ell'} R_{n\ell} \, dr \right] \left[\int_{-1}^{+1} \mu P_{\ell'}^{|m'|}(\mu) P_\ell^{|m|}(\mu) \, d\mu \right]$$

$$\times \left[\int_0^{2\pi} e^{i(m-m')\phi} \, d\phi \right]$$

where we have expressed the eigenfunctions explicitly. In order for the integral over ϕ not to vanish, $\Rightarrow m = m'$. Now consider the recursion relation

$$\mu P_\ell^{(m)}(\mu) = \frac{1}{2\ell + 1} \left[(\ell - |m|) \, P_{\ell+1}^{|m|}(\mu) + (\ell + |m|) \, P_{\ell-1}^{|m|}(\mu) \right]$$

and substitute into the μ integral, above, and use the orthogonality relationship of associated Legendre polynomials. The condition for the θ integral not to vanish is then given by

$$\left. \begin{array}{l} \ell' = \ell + 1 \\ \ell' = \ell - 1 \end{array} \right\} \Delta\ell = \pm 1.$$

Thus

$$\langle \phi_f \, |z| \, \phi_i \rangle = \left\{ \begin{array}{l} 0, \ \Delta\ell \neq \pm 1, \ \Delta m \neq 0 \\ > 0, \ \Delta\ell = \pm 1, \ \Delta m = 0. \end{array} \right.$$

Similar consideration of $\langle \phi_f \, |x \pm iy| \, \phi_i \rangle$ yields

$$\langle \phi_f \, |x \pm iy| \, \phi_i \rangle = \left\{ \begin{array}{l} 0, \ \Delta\ell \neq \pm 1, \ \Delta m \neq \pm 1 \\ > 0, \ \Delta\ell = \pm 1, \ \Delta m = \pm 1. \end{array} \right.$$

These electric dipole selection rules limit the amount of momentum the photon can take away.

12.8.2 *Electric quadrupole transitions*

Transitions whose dipole moments are zero are said to be forbidden. Consideration of higher order terms of the multipole expansion, however, indicates that this is true only in the dipole approximation. The higher terms, such as the electric quadrupole, can yield finite, though usually much lower, transition probabilities. Recall that the general expression for the transition matrix element is given by

$$\langle \phi_f | e^{i\vec{k}\cdot\vec{x}} \vec{p} | \phi_i \rangle \cdot \hat{e}_\alpha$$

where

$$e^{i\vec{k}\cdot\vec{x}} = 1 + i\vec{k}\cdot\vec{x} + \cdots$$

For the next higher term, we must evaluate

$$i\vec{k} \cdot \langle \phi_f | \vec{x}\vec{p} | \phi_i \rangle \cdot \hat{e}_\alpha.$$
$$\uparrow \text{dyad}$$

Proceed as with classical multipoles

$$\vec{x}\vec{p} = \frac{1}{2}(\vec{x}\vec{p} + \vec{p}\vec{x}) + \frac{1}{2}(\vec{x}\vec{p} - \vec{p}\vec{x}).$$

Then note the following identities

$$\vec{x}\vec{p} + \vec{p}\vec{x} = \frac{im_e}{\hbar}[H_0, \vec{x}\vec{x}]$$

$$\vec{k} \cdot (\vec{x}\vec{p} - \vec{p}\vec{x}) \cdot \hat{e}_\alpha = (\vec{k} \times \hat{e}_\alpha) \cdot (\vec{x} \times \vec{p}).$$

Given that

$$\vec{k} \cdot \hat{e}_\alpha = 0$$

$$\Rightarrow \vec{k} \cdot (\vec{x}\vec{x}) \cdot \hat{e}_\alpha = \vec{k} \cdot \left(\vec{x}\vec{x} - \frac{|\vec{x}|^2}{3} \tilde{I} \right) \cdot \hat{e}_\alpha$$

$$= (1/3e)\vec{k} \cdot (-e[3\vec{x}\vec{x} - |\vec{x}|^2\tilde{I}]) \cdot \hat{e}_\alpha$$

$$= (1/3e)\vec{k} \cdot \tilde{Q} \cdot \hat{e}_\alpha$$

where

$$\tilde{Q} \equiv -e(3\vec{x}\vec{x} - |\vec{x}|^2\tilde{I})$$

$$Q_{33} = -e(2z^2 - x^2 - y^2) = (-er^2)2(4\pi/5)^{1/2}Y_{2,0}$$

$$Q_{31} \pm i Q_{23} = -e3z(x \pm iy) = (\pm er^2)3(8\pi/15)^{1/2}Y_{2,\pm1}$$

$$Q_{11} \pm 2i Q_{12} - Q_{22} = -e3(x \pm iy)^2 = (-er^2)12(2\pi/15)^{1/2}Y_{2,\pm2}$$

are sufficient to specify a traceless, symmetric $\tilde{Q}$. The selection rules that follow are governed by $\Delta \ell = 0, \pm 2$, $\Delta m = 0, \pm 1, \pm 2$.

To calculate the transition matrix elements, we also need to evaluate

$$\oint Y^*_{\ell_3 m_3} Y_{\ell_2 m_2} Y_{\ell_1 m_1} d\Omega = \left[\frac{(2\ell_1 + 1)(2\ell_2 + 1)}{4\pi (2\ell_3 + 1)} \right]^{1/2}$$

$$\times \langle \ell_1 \ell_2 00 | \ell_1 \ell_2 \ell_3 0 \rangle \langle \ell_1 \ell_2 m_1 m_2 | \ell_1 \ell_2 \ell_3 m_3 \rangle .$$

The symbols inside the angle brackets are known as the Clebsch–Gordon coefficients.

12.9   Numerical evaluation of transition probabilities

As an illustration of how we calculate dipole transition probabilities and absorption cross-sections we will calculate some numbers for the $2 \leftrightarrow 1$ transition of hydrogen, the so-called Lyman α transition. We will begin in general terms by considering any downward transition, $(n', l', m') \to (n, l, m)$ where $n' > n$.

The essential computation is the calculation of the dipole matrix element, $\langle i | \vec{x} | f \rangle$, where

$$|nlm\rangle = R_{nl}(r) Y_{lm}(\theta, \phi).$$

Recall that for the H atom

$$R_{nl}(r) = \left\{ \frac{(n-l-1)!}{2n \left[(n+l)!\right]^3} \left(\frac{2}{na_0}\right)^3 \right\}^{1/2} e^{-r/na_0} \left(\frac{2r}{na_0}\right)^l L_{n+1}^{2l+1}\left(\frac{2r}{na_0}\right)$$

and

$$Y_{lm}(\theta, \phi) = \left[\frac{(l-|m|)!(2l+1)}{4\pi(l+|m|)!} \right]^{1/2} P_l^{|m|}(\cos \theta) \, e^{im\phi}.$$

The vector, $\vec{x} = x\hat{e}_x + y\hat{e}_y + z\hat{e}_z$, can be cast into the form

$$\vec{x} = \frac{1}{2}[(x+iy)(\hat{e}_x - i\hat{e}_y) + (x-iy)(\hat{e}_x + i\hat{e}_y) + z\hat{e}_z].$$

This format has a useful physical interpretation. The case $\hat{e}_\alpha \cdot \vec{x} = z$ represents interaction with a wave polarized in the z direction (and therefore propagating in the x–y plane). The case $\hat{e}_\alpha \cdot \vec{x} \propto x \pm iy$ represents a wave traveling in the z direction and having right or left circular polarization.

The polarization directions can be expressed in terms of Y_{lm} so that

$$z = r \cos \theta = \left(\frac{4\pi}{3}\right)^{1/2} r Y_{10}$$

and

$$x \pm iy = r \sin\theta \, e^{\pm i\phi} = \left(\frac{8\pi}{3}\right)^{1/2} rY_{1,\pm 1}.$$

Now we are ready to evaluate the square of the dipole matrix element, $|\langle n'l'm'|\vec{x}|nlm\rangle|^2$.

Substituting all of this, and rearranging terms, we get

$$\left|\langle n'l'm'|\vec{x}|nlm\rangle\right|^2 = a_0^2 \mathcal{Y}_{l'm'lm} \mathcal{R}_{n'l'nl}^2$$

where

$$\mathcal{Y}_{l'm'lm} \equiv \frac{4\pi}{3}\left[\left|\oint Y^*_{l'm'} Y_{11} Y_{lm}\, d\Omega\right|^2 + \left|\oint Y^*_{l'm'} Y_{10} Y_{lm}\, d\Omega\right|^2 \right.$$
$$\left. + \left|\oint Y^*_{l'm'} Y_{1,-1} Y_{lm}\, d\Omega\right|^2\right]$$

and

$$\mathcal{R}_{n'l'nl} \equiv \frac{1}{a_0}\int_0^\infty R_{n'l'} R_{nl}\, r^3\, dr.$$

The volume integral was separated according to $dV = dA\, dr = r^2\, d\Omega\, dr$.

12.9.1 The Lyman α transition

Let us now evaluate the specific transition $(2, 1, m') \rightarrow (1, 0, 0)$. Recall that there are three possible values of m' because the range in m is given by $2l + 1$. Thus for, $l = 1, m = -1, 0, 1$. For each of these cases, $\mathcal{Y}$ must be evaluated separately. As it turns out the $\mathcal{Y}$'s are identical and equal to $1/3$ (check it out yourself). Thus

$$\mathcal{Y}_{1,1,0,0} = \mathcal{Y}_{1,0,0,0} = \mathcal{Y}_{1,-1,0,0} = 1/3.$$

In the case of the radial wave functions

$$R_{1,0}(r) = 2a_0^{-3/2}\, e^{-r/a_0}$$

$$R_{2,1}(r) = (24)^{-1/2} a_0^{-3/2}\left(\frac{r}{a_0}\right) e^{-r/2a_0}$$

so that

$$\mathcal{R}_{2,1,1,0} = \frac{1}{a_0^4}\int_0^\infty 6^{-1/2} r^4\, e^{-3r/2a_0}\, dr.$$

Now, let us change variables so that $x = r/a_0$, and

$$\mathcal{R}_{2,1,1,0} = \frac{1}{\sqrt{6}} \int_0^\infty x^4 e^{-\frac{3}{2}x} \, dx = \frac{1}{\sqrt{6}} \left(\frac{2}{3}\right)^5 4!$$

Substituting into the above yields

$$|\langle 211|\vec{x}|100\rangle|^2 = |\langle 210|\vec{x}|100\rangle|^2 = |\langle 21, -1|\vec{x}|100\rangle|^2 = \frac{2^{15}}{3^{10}} a_0^2.$$

Taking $a_0 = $ Bohr radius $= 5.29 \times 10^{-9}$ cm, we finally have

$$|\langle 21m'|\vec{x}|100\rangle|^2 = 1.55 \times 10^{-17}.$$

Substituting into (54) now yields the transition rate

$$A_{2p1s} = 1.55 \times 10^{-17} \frac{4e^2}{3\hbar} \left(\frac{\omega_{21}}{c}\right)^3 \text{ s}^{-1}.$$

In the case of the Lyman α (Ly-α) transition

$$\hbar\omega_{21} = 13.6 \times 1.6 \times 10^{-12} \left(\frac{1}{n^2} - \frac{1}{n'^2}\right) \rightarrow \omega_{21} = 1.6 \times 10^{16} \text{ s}^{-1}.$$

Thus

$$A_{2p1s} = 6.8 \times 10^8 \text{ s}^{-1}.$$

Is this the actual transition rate? To answer that question we must ask how many possible states the electron could occupy in the $n = 2$ level. It can occupy the three degenerate states already described but it can also occupy the 2s state $(2, 0, 0)$.

Effective transition rate

To get the effective transition rate we must average over all possible transition rates for all possible state changes in the $n' \rightarrow n$ transition. Generally, since level n has m^2 possible states, we have

$$A_{n'n} = \sum_{l=0}^{n-1} \sum_{l'=0}^{n'-1} \frac{2l' + 1}{n'^2} A_{n'l'nl}.$$

In the case of the Ly-α transition

$$A_{21} = \frac{1}{4}[3A_{2110} + A_{2010}] = \frac{1}{4}[3A_{2p1s} + A_{2s1s}].$$

Thus, we must evaluate the A_{2s1s} transition.

If we proceed as before we quickly find that

$$\mathcal{Y}_{0000} \propto Y_{00}^* Y_{11} Y_{00} + Y_{00}^* Y_{10} Y_{00} + Y_{00}^* Y_{1,-1} Y_{00} = 0$$

so that, $A_{2s1s} = 0$, consistent with the dipole selection rules.

As we have seen, there are three states corresponding to A_{2p1s} whereas there is only one state corresponding to A_{2s1s}. Thus

$$A_{21} = \frac{3}{4} A_{2p1s} = 5 \times 10^8 \text{ s}^{-1}.$$

Cross-section for Ly-α absorption

We can now calculate the cross-section for Ly-α absorption. At the line center

$$\mathcal{L}(\omega = \omega_{12}) = \frac{2}{\pi \Gamma} = \frac{2}{A_{21}}$$

so that

$$\sigma_{12}(\omega = \omega_{12}) = \frac{8\pi}{3} \left(\frac{e^2}{\hbar c} \right) \omega_{12} |\vec{X}_{12}|^2 / A_{21}.$$

Using these numbers we get

$$\sigma_{12}(\omega = \omega_{12}) = 3 \times 10^{-11} \text{ cm}^2.$$

The Ly-α absorption rate

The relationship between photon occupation number, $N_\alpha(\omega)$, and the Planck curve in the case of black-body radiation is given by

$$I(\omega) = \sum_1^2 \frac{\hbar \omega^3}{(2\pi)^3 c^2} N_\alpha(\omega).$$

The transition rate is defined by

$$\frac{\partial P_{12}}{\partial t} = B_{12} I(\omega = \omega_{12})$$

so that

$$B_{12} = \frac{2e^2}{3(2\pi)^2 c} |\vec{X}_{12}|^2 = \frac{(2\pi)^2 c^2}{2\hbar \omega^3} A_{21}.$$

Numerically, the above yields

$$B_{12} = 8.3 A_{21} = 4.2 \times 10^9 \text{ s}^{-1}.$$

12.9.2 Bound–free absorption cross-section

Substituting the physical constants into (341) yields

$$\sigma_{bf} = 0.7 \times 10^{32} \omega^{-3}$$

for the H atom.

A photon which can just ionize the H atom has $\hbar\omega = 13.6$ eV $\rightarrow \omega = 2 \times 10^{16}$ s^{-1} so that

$$\sigma_{1f} = 9 \times 10^{-18} \text{ cm}^2.$$

A 100 eV, soft X-ray photon has $\omega = 1.5 \times 10^{17}$ s^{-1} yielding

$$\sigma_{1f} = 2 \times 10^{-20} \text{ cm}^2.$$

A 5 keV, hard X-ray photon has $\omega = 7.5 \times 10^{18}$ s^{-1} yielding

$$\sigma_{1f} = 1.5 \times 10^{-25} \text{ cm}^2.$$

We see that soft X-rays are attenuated much more than hard X-rays as the photons propagate through the interstellar medium.

12.10   HII regions

We will now apply some of what we have learned thus far to examine the interesting effects associated with the interaction of a star with a surrounding nebula. Probably the most famous example is the Orion nebula (Fig. 12.4). We begin by examining the properties of a star that lead to such an interaction. Then, we will examine the ionization process that leads to the formation of an HII region. Finally, the properties of the HII regions are discussed.

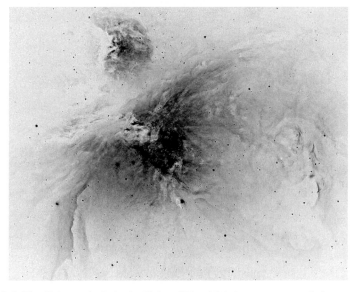

Fig. 12.4 The Orion nebula in the light of H α. This image was made by recording the H α photons emitted by the photo-ionized nebula.

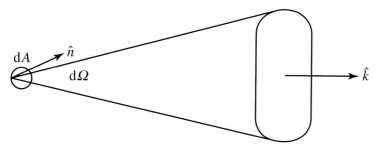

Fig. 12.5 The geometry that defines the intensity,

12.10.1 Ionizing stars

Stars with photospheric temperatures in excess of 10^4 K emit a significant number of photons whose energies are greater than 13.6 eV, the ionization potential of hydrogen. Consequently, if such a star is immersed in a cloud of neutral hydrogen, some of that cloud will be ionized, leading to the formation of an HII region, that is, a region of ionized hydrogen.

The spectrum of a star, to a good approximation, is that of a black body whose temperature is that of the star's photosphere. The spectrum can therefore be described with the Planck law

$$I_\nu = \frac{2h\nu^3/c^2}{e^{h\nu/kT} + 1} \; \text{erg cm}^{-2}\,\text{s}^{-1}\,\text{Hz}^{-1}\,\text{sr}^{-1}$$

where ν is the frequency of the emitted radiation and T is the temperature of the photosphere. The intensity, I_ν is defined by

$$dE = I_\nu \hat{k} \cdot \hat{n} \; dA \; d\Omega \; d\nu \; dt$$

where $\hat{k}$ is the direction of the emitted radiation and $\hat{n}$ is the direction of the normal of the surface from which the radiation is emitted (see Fig. 12.5).

In order for a star to ionize HI, it must produce reasonable quantities of Lyman continuum photons (photons with $h\nu > 13.6\,\text{eV}$ or $\nu > \nu_0 = 3 \times 10^{15}$ Hz). In terms of photons, the Planck curve can be rewritten as

$$\dot{\mathcal{N}}_\epsilon = 1.6 \times 10^{22} \frac{2\epsilon^2}{e^{\epsilon/kT} + 1} \; \text{cm}^{-2}\,\text{s}^{-1}\,\text{eV}^{-1}\,\text{sr}^{-1} \qquad (361)$$

where ϵ is the energy of the photon in electronvolts. Thus the number of photons emitted with $\epsilon > 13.6$ eV is given by

$$\dot{\mathcal{N}}_i = 1.6 \times 10^{22} \int_{13.6\,\text{eV}}^{\infty} \frac{2\epsilon^2}{e^{\epsilon/kT} + 1} \; d\epsilon \; \text{cm}^{-2}\,\text{s}^{-1}\,\text{sr}^{-1}.$$

The fraction of a star's emitted photons that can ionize HI is therefore given by

$$f_i = \int_{h\nu_0}^{\infty} \frac{2\epsilon^2}{e^{\epsilon/kT}+1} \bigg/ \int_0^{\infty} \frac{2\epsilon^2}{e^{\epsilon/kT}+1} = \dot{N}_i/\dot{N}.$$

For the Sun, which has a photospheric temperature of 5600 K, the fraction is 4×10^{-10}. For an A0 star with a temperature of 10 000 K, the fraction is 3×10^{-5}. For a B0 star with $T = 30\,000$ K, 12% of the emitted photons are ionizing. In the case of the hottest stars, the O5 stars, the fraction is more like 45%.

Integration of (361) with respect to dA, $d\Omega$ and $d\epsilon$ yields the luminosity of the star in photons per second. For a spherically symmetric star

$$\dot{N} = (4\pi)^2 \dot{N} R_*^2$$

and

$$\dot{N}_i = (4\pi)^2 \dot{N}_i R_*^2 = 10^{46} f_i \, (R_*/R_\odot)^2 \, \text{s}^{-1}$$

where R_* is the radius of the star.

12.11   Ionization of a pure hydrogen nebula

The ionizing radiation of the star interacts with the surrounding HI atoms. The number of such photons incident on the atoms per unit area per unit time is obtained by integrating (361) with respect to the solid angle, $d\Omega$. The ionization rate per unit volume is then given by the product of the atom number density, the ionization rate and the cross-section for bound–free absorption

$$1.6 \times 10^{22} 4\pi \, n_H \int_{13.6\,\text{eV}}^{\infty} \frac{2\epsilon^2 \, \sigma_{bf}(\epsilon)}{e^{\epsilon/kT}+1} \, d\epsilon \; \text{cm}^{-3} \, \text{s}^{-1}.$$

The cross-section can be approximated by recognizing that almost all of the H atoms are in the ground state when neutral. One way to see that is to consider the spontaneous transition probabilities for hydrogen. Some of the various transitions are shown in Fig. 12.6.

The transition rate between any two levels can be approximated by the formula

$$A_{nn'} \approx 1.6 \times 10^{10} [n^3 n' (n^2 - n'^2)]^{-1}.$$

Some examples

$$A_{2,1} = 5 \times 10^8 \, \text{s}^{-1}$$

$$A_{5,4} = 3.6 \times 10^6 \, \text{s}^{-1}$$

$$A_{10,9} = 9.3 \times 10^4 \, \text{s}^{-1}$$

$$A_{110,109} = 0.5 \, \text{s}^{-1}.$$

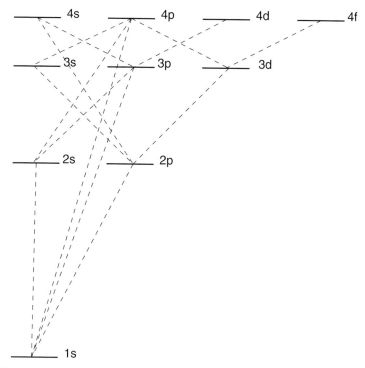

Fig. 12.6 The energy level diagram for hydrogen. Energy levels to $n = 4$ are shown. The dashed lines indicate all permitted transitions to $n = 3$ and lower.

We see that there is a strong tendency for the electron to descend into the ground state. Thus, the approximation that all electrons are in the ground level is a reasonable one. We therefore make the approximation that $\sigma_{bf}(v) = \sigma_{1f}(v) = \sigma_v$ so that (360) can be used. The electrons freed by the ionizations will recombine at a rate given by

$$n_p n_e \alpha_A$$

where $\alpha_A \propto v_e \sigma_{fb}$ is the recombination coefficient and has units cm^3 s^{-1}. It is defined by

$$\alpha_A = \sum_{n,l} \alpha_{nl}$$

and

$$\alpha_{n,l} = \int_0^\infty v \sigma_{nl} f(v) \, dv$$

where

$$f(v) = \frac{4}{\sqrt{\pi}} \left(\frac{m}{2kT} \right)^{3/2} v^2 \, e^{-mv^2/2kT}.$$

See tabulated values of α in Table 12.1.

Table 12.1. *Recombination coefficients for $T = 10^4$ K in units of $10^{-15} \text{cm}^3 \text{ s}^{-1}$*

α_{1s}	α_{2s}	α_{2p}	α_{3s}	α_{3p}	α_{3d}	α_{4s}	α_{4p}	α_{4d}	α_{4f}	$\rightarrow$	α_A	α_B
160	23	54	7.8	20	17	3.6	9.7	11	5.5	$\rightarrow$	420	260

An equilibrium is established so that the ionization rate equals the recombination rate

$$1.6 \times 10^{22} 4\pi n_H \int_{13.6 \text{ eV}}^{\infty} \frac{2\epsilon^2 \sigma_{bf}(\epsilon)}{e^{\epsilon/kT} + 1} \, d\epsilon = n_p n_e \alpha_A.$$

The ionizing photons are first absorbed by the H atoms. The subsequent recombinations can produce additional ionizing photons which in turn are absorbed. Thus, as we proceed away from the star, the local radiation field will be determined by a combination of diluted star light and local emission from the atoms that are recombining. The intensity of the radiation as a function of distance from the star is therefore defined by a differential equation called "the equation of radiative transfer" which can be written as

$$\frac{dI_\nu}{dr} = -n_H \sigma_\nu I_\nu + j_\nu$$

where the two terms on the right-hand side represent absorption and emission per unit volume respectively, and where $I_\nu = I_{\nu s} + I_{\nu d}$ represent the stellar and diffuse (from the gas) radiation. Thus, for the stellar field alone

$$I_{\nu s}(r) = I_{\nu s}(R_*) \, e^{-\tau_\nu}$$

where

$$\tau_\nu(r) = \int_0^r n_H(r') \sigma_\nu \, dr'. \tag{362}$$

For the *diffuse* radiation field we must include a source term (for the local recombination), so that the equation of transfer is

$$\frac{dI_{\nu d}}{dr} = -n_H \sigma_\nu I_{\nu d} + j_\nu$$

where

$$j_\nu = \frac{2h\nu^3}{c^2} \left(\frac{h^2}{2\pi mkT} \right)^{3/2} e^{-h(\nu - \nu_0)/kT} n_p n_e \qquad (\nu > \nu_0)$$

(see Osterbrock, 1974). Since every photon created by a recombination to the ground level is an ionizing photon it will eventually be absorbed (as long as there is enough

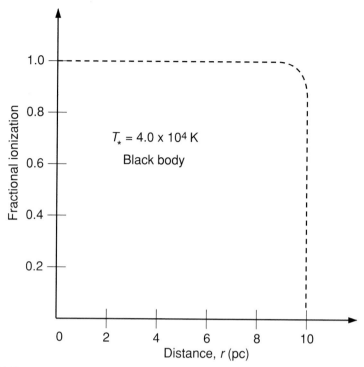

Fig. 12.7 Fractional ionization as a function of distance from the center of an HII region. Note the sharp fall-off that defines the edge of the HII region.

gas around). Thus, any such photons must be local in origin so that $dI_{vd}/dr = 0$ and

$$I_{vd} = \frac{j_v}{n_H \sigma_v}$$

and

$$4\pi \int_{v_0}^{\infty} \frac{j_v}{h v} dv = n_p n_e \alpha_1.$$

Thus, for any point in the nebula

$$4\pi n_H \int_{v_0}^{\infty} \frac{I_v}{h v} \sigma_v \, e^{-\tau_v(r)} \, dv + n_p n_e \alpha_1 = n_p n_e \alpha_A.$$

But $\alpha_A = \alpha_B + \alpha_1$, so we finally have

$$4\pi n_H \int_{v_0}^{\infty} \frac{I_v}{h v} \sigma_v \, e^{-\tau_v(r)} \, dv = n_p n_e \alpha_B. \tag{363}$$

Equations (362) and (363) can be used to solve for $n_H(r)$ and $n_e(r)$ given an input gas density distribution, $n(r) = n_H(r) + n_e(r)$. A typical solution is shown in Fig. 12.7.

Note that the HII region has a very sharp boundary. The gas is fully ionized inside the boundary and fully neutral outside.

12.11.1 Radius of HII region

Equations (362) and (363) can also be combined to solve for the radius of the HII region. Taking the differential form of (362)

$$\frac{d\tau_\nu}{dr} = n_H \sigma_\nu.$$

Combining with (363) yields

$$4\pi \int_{\nu_0}^\infty \frac{I_\nu}{h\nu} d\nu \frac{d\tau}{dr} e^{-\tau_\nu(r)} = n_p n_e \alpha_B.$$

Integrating over all space and assuming spherical symmetry

$$4\pi \int_{\nu_0}^\infty \frac{I_\nu}{h\nu} d\nu \int_0^\infty d(-e^{-\tau_\nu})4\pi r^2 \, dr = \int_0^{r_0} n_p n_e \alpha_B 4\pi r^2 \, dr.$$

On the right-hand side we have recognized that there is no ionized gas outside r_0.
Evaluating these integrals and rearranging terms, we have

$$\int_{\nu_0}^\infty (4\pi)^2 r^2 I_\nu / h\nu \, d\nu = \frac{4}{3}\pi r_0^3 n_p n_e \alpha_B.$$

But $L_\nu = (4\pi)^2 r^2 I_\nu$, so that

$$\int_{\nu_0}^\infty L_\nu / h\nu \, d\nu = \dot{N}_i = \frac{4}{3}\pi n_p n_e \alpha_B r_0^3.$$

Solving this for r_0

$$r_0 = \left[\frac{3\dot{N}_i}{4\pi n_e^2 \alpha_B}\right]^{1/3} = 10^4 \dot{N}_i^{1/3} \left(\frac{n_e}{1 \text{ cm}^{-3}}\right)^{-2/3}. \tag{364}$$

The quantity r_0 is often referred to as the Stromgren radius and the quantity $(4/3)\pi r_0^3$ as the Stromgren sphere.

According to (364), $\dot{N}_i$ can be calculated for any star. Table 12.2 lists the size of the Stromgren sphere predicted for four different types of star.

12.12 Quasars and the Lyman α forest

Quasars are compact, stellar-like objects when imaged on photographic plates or by CCD cameras. What makes them unique is their often tremendous redshifts. Figure 12.8 shows the spectra of four distant quasars.

Table 12.2. *Stromgren spheres. The table shows the size of the ionizing star in solar radii, the log of the ionization rate and the size of the corresponding HII region in parsecs. An ambient density of one particle per cubic centimeter is assumed.*

Spectral type	$R/R_\odot$	$\log \dot{N}_i$	r_0 (pc)
O5	10	50.0	142
B0	7.0	48.7	50
A0	2.5	42.7	0.6
G2	1.0	36.3	5×10^{-3}

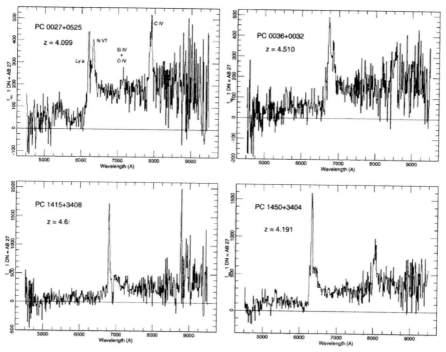

Fig. 12.8 Spectra of four quasars with $z > 4$. The spectral lines are annotated. Note the observed wavelengths of the L_y-α line compared to the rest wavelength of 1216Å. From Schneider, D. P., Schmidt, M. and Gunn, J. E. 1997, AJ, 114, pp. 36–40.

The most distant quasars also show complex absorption spectra which contain the same line observed over a wide range of wavelengths. For example, Fig. 12.9 shows a large number of Lyman α lines spaced out over a wide range of wavelengths. It is doubtful that a single source could account for the range in velocity (and therefore distance). It has therefore been proposed that the absorption line system (the forest) is caused by intervening galaxies or clouds of gas. The intervening objects, although

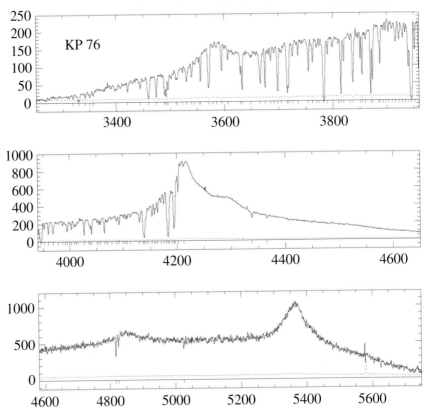

Fig. 12.9 Lyman α forest. The observations are taken from Crotts, A. P. S. and Fang, Y. (1998, ApJ, 502, pp. 16–47).

closer than the quasar in question, are probably also in their earlier phases of evolution. The distribution of these objects in velocity space provides an important clue on the distribution of matter in the distant Universe. This is particularly true given that the matter sampled in the Lyman α forest is fairly old, that is the Lyman α photons left the scene billions of years ago. Did the Universe at that time have similar structure to the modern Universe?

12.12.1 *Correlation studies*

Features in the Lyman α forest are separated in velocity space by amounts ranging from the instrumental spectral resolution to the entire range in velocities observed. The question is whether the distribution of possible values of Δv is different from a random distribution.

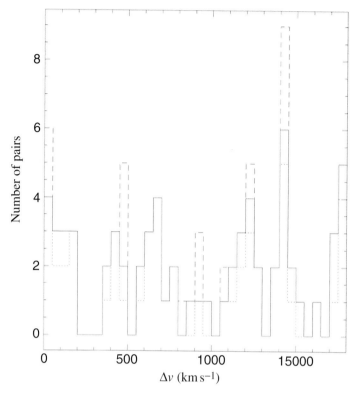

Fig. 12.10 Correlation statistics for a Lyman α forest. The figure is taken from Crotts, A. P. S. and Fang, Y. (1998, ApJ, 502, pp. 16–47).

Let

$$\Delta v = \frac{c\left[\left(\frac{\Delta\lambda}{\lambda_0}\right)_1 - \left(\frac{\Delta\lambda}{\lambda_0}\right)_2\right]}{1 + \left(\frac{\Delta\lambda}{\lambda_0}\right)}$$

where the subscripts 1 and 2 refer to any pair of Lyman α lines. We can then define a two-point correlation function, ζ, such that

$$\zeta(\Delta v) = \frac{N_{obs}(\Delta v)}{N_p(\Delta v)} - 1$$

where $N_{obs}(\Delta v)$ is the number of observed pairs at a given Δv and $N_p(\Delta v)$ is the number expected from a purely random distribution.

Figure 12.10 is taken from Crotts and Fang (1998) and shows a correlation plot. Note that the excursions from $\zeta = 0$ are not significant, suggesting that there is no strong clustering of Lyman α lines in the forest. Such results suggest that if the lines arise from intervening galaxies, the galaxies are not as strongly clustered as they have become. Some astronomers have taken this to mean that the intervening matter is not galaxies at all but rather a general intergalactic medium in the form of clouds. The issue is not yet totally settled. If there is an intergalactic medium it is fairly tenuous as the following estimate shows.

12.12.2 Column density of the HI responsible for the $Ly - \alpha$ forest

Using our definition of optical depth we write down the opacity arising from one cloud as

$$\delta\tau = n_1\sigma_{12}\delta r = dN_1\sigma_{12}$$

where n_1 represents the number density of hydrogen atoms in the ground state, $\sigma_{12} = 3 \times 10^{-11}$ cm^2 is the Lyman α absorption cross-section as defined on page 209 and δr is taken to be the dimension of the cloud responsible for a specific Lyman α feature. The column density associated with one cloud is δN_1.

If the clouds are moving as a result of the Hubble flow then we can relate δr to the line spacings in the Ly-α forest

$$dv = H \, \delta r$$

where dv is the minimum resolved line spacing in velocity space.

The observed absorption lines are unsaturated suggesting that $\delta\tau < 1$ so that

$$dN_1\sigma_{12} \leq 1 \rightarrow dN_1 \leq 3 \times 10^{10} \text{ cm}^{-2}.$$

Summing over the entire Lyman α forest

$$N_H = \int_0^{r_0} n_1 \, \delta r = \left(\frac{dN_1}{\delta r}\right) r_0 \approx \langle dN_1 \rangle \frac{v_0}{dv}$$

where v_0 is the recession velocity of the quasar. Using these numbers

$$N_H \leq 3 \times 10^{10} \frac{2.7 \times 10^{10}}{5 \times 10^6} \approx 2 \times 10^{14} \text{ cm}^{-2}$$

which is about six orders of magnitude smaller than the column density of HI in our own Galaxy.

12.13 Reference

Crotts, A. P. S. and Fang, Y. (1998) Re-observation of Close QSO Groups: The Size Evolution and Shape of Ly alpha Forest Absorbers, *The Astrophysical Journal*, **502**, pp. 16–47.

12.14 Further reading

Cohen-Tanoudji, C. (1992) *Principles of Quantum Mechanics*, John Wiley and Sons, New York, NY, USA.

Griffiths, D. (1995) *Introduction to Quantum Mechanics*, Prentice Hall, Englewood Cliffs, NJ, USA.

Landau, L. D. and Lifschitz, E. M. (1977) *Quantum Mechanics*, 3rd edn, Pergamon Press, Oxford, UK.

Landau, R. H. (1996) *Quantum Mechanics II: A Second Course in Quantum Theory*, 2nd edn, John Wiley and Sons, New York, NY, USA.

Osterbrock, D. E. (1974) *Astrophysics of Gaseous Nebulae*, W. H. Freeman, San Francisco, CA, USA.

Rybicki, G. B. and Lightman, A. P. (1985) *Radiative Processes in Astrophysics*, John Wiley and Sons, New York, NY, USA.

Shu, F. H. (1992) *The Physics of Astrophysics: Radiation*, University Science Books, Mill Valley, CA, USA.

Chapter 13

Atomic fine structure lines

13.1 Electron spin

Some of the most important astrophysical transitions involve electron spin. Perhaps the most famous example is the 21 cm line. We will now consider the role of electron spin in atomic transitions. The non-relativistic Hamiltonian, introduced earlier, contains no spin-specific terms (it allows only for orbital angular momentum through the operator L). As I will now demonstrate, the relativistic Hamiltonian provides additional degrees of freedom so that effects like spin (nuclear and electron) can be incorporated into the Schrödinger equation. To that end we now discuss Dirac's postulate, spin angular momentum, Dirac's equation, the non-relativistic limit, relativistic corrections, transitions involving spin, the Zeeman effect and spin–orbit coupling. Discussion of these topics will give us an understanding of atomic fine structure.

13.1.1 Relativistic Hamiltonian

Recall the form of the relativistic Hamiltonian (see, for example, Shu, 1992).

$$H = [|c\vec{P}|^2 + m^2 c^4]^{1/2} + q\Phi \tag{365}$$

where

$$\vec{P} = \vec{p} - \frac{q}{c}\vec{A}.$$

It is difficult to use the above as an operator because it is nonlinear. Let us therefore square the operators of

$$H\Psi = i\hbar \frac{\partial \Psi}{\partial t} \rightarrow H^2 \Psi = -\hbar \frac{\partial^2}{\partial t^2}\Psi.$$

With no *scalar* potential the above reduces to

$$\left[(|c\vec{P}|^2 + m^2c^4) - \hbar^2\frac{\partial^2}{\partial t^2}\right]\Psi = 0.$$

With no vector potential we get the Klein–Gordon (K–G) equation

$$\left[\left(\nabla^2 - \frac{1}{c^2}\frac{\partial^2}{\partial t^2}\right) - \left(\frac{mc}{\hbar}\right)^2\right]\Psi = 0. \tag{366}$$

Note that Ψ is not a wave function. It is a scalar field which must be described using Quantum Field theory.

13.2 Dirac's postulate

To circumvent the problem with the K–G equation, Dirac proposed a Hamiltonian of the form

$$H = \vec{a} \cdot \vec{P}_a + bmc^2 + q\phi \tag{367}$$

where $\vec{a}$ and b are constants.

To make (367) equivalent to (365) requires

$$a_x a_x = a_y a_y = a_z a_z = bb = 1$$
$$a_x a_y + a_y a_x = a_y a_z + a_z a_y = a_z a_x + a_x a_z$$
$$= a_x b + b a_x = a_y b + b a_y = a_z b + b a_z = 0.$$

This can only hold if a_x, a_y, a_z and b form 4×4 matrices. Thus

$$a_x = \begin{pmatrix} O & \sigma_x \\ \sigma_x & O \end{pmatrix} \quad \text{and} \quad a_y = \begin{pmatrix} O & \sigma_y \\ \sigma_y & O \end{pmatrix}$$

$$a_z = \begin{pmatrix} O & \sigma_z \\ \sigma_z & O \end{pmatrix} \quad \text{and} \quad b = \begin{pmatrix} I & O \\ O & -I \end{pmatrix} \tag{368}$$

where

$$O \equiv \begin{pmatrix} O & O \\ O & O \end{pmatrix} \quad \text{and} \quad I \equiv \begin{pmatrix} 1 & O \\ O & 1 \end{pmatrix}$$

and the Pauli matrices are

$$\sigma_x = \begin{pmatrix} O & 1 \\ 1 & O \end{pmatrix} \quad \text{and} \quad \sigma_y = \begin{pmatrix} O & -i \\ i & O \end{pmatrix} \quad \text{and} \quad \sigma_z = \begin{pmatrix} 1 & O \\ O & -1 \end{pmatrix}. \tag{369}$$

See Shu (pp. 268–9 in that book) for suggestions on proof. The properties of the Pauli matrices can be summarized as

$$\sigma_i \sigma_k - \sigma_k \sigma_i = 2i\epsilon_{ikm}\sigma_m$$

$$\sigma_i \sigma_k + \sigma_k \sigma_i = 2\delta_{ik}. \tag{370}$$

Rotation and angular momentum

We can write (370) in operator form, so that

$$\vec{\sigma} \times \vec{\sigma} = 2i\vec{\sigma}.$$

Note the analogy to

$$\vec{L} \times \vec{L} = i\hbar\vec{L}$$

$$\Rightarrow \text{let} \quad \vec{s} = \frac{\hbar}{2}\vec{\sigma} \Rightarrow \vec{s} \times \vec{s} = i\hbar\vec{s}.$$

13.2.1 The Dirac equation

The Dirac equation gives us a mathematical feel for the extra degrees of freedom that allow us to include spin in the definition of the wave function, Ψ.

To make use of (367) we need a four-component Ψ because the operators are 4×4 matrices,

$$\Psi \Rightarrow \chi \equiv \begin{pmatrix} \chi_1 \\ \chi_2 \\ \chi_3 \\ \chi_4 \end{pmatrix}$$

so that

$$\Rightarrow \left[c\vec{\alpha} \cdot \left(i\hbar\vec{\nabla} + \frac{e}{c}\vec{A}\right) + bm_ec^2\right]\chi = \left(i\hbar\frac{\partial}{\partial t} + e\phi\right)\chi. \tag{371}$$

It follows that the probability density is given by

$$g = \chi^\dagger \chi = |\chi_1|^2 + |\chi_2|^2 + |\chi_3|^2 + |\chi_4|^2$$

where $\dagger$ = Hermitian complex conjugate.

13.2.2 Free particle at rest

Normally, the Dirac equation is too complicated to be solved analytically. However, there exists a simple but important application of the Dirac equation. It will serve to illustrate how the Dirac equation is used in practice.

For a free particle, ϕ, $\vec{A} = 0$, so that the Dirac equation becomes

$$\Rightarrow [-i\hbar c\vec{\alpha} \cdot \vec{\nabla} + bm_e c^2]\chi = i\hbar \frac{\partial \chi}{\partial t}. \tag{372}$$

Let us try a solution of the form

$$\chi = \chi_0(\vec{k}_e) \, e^{i(\vec{k}_e \cdot \vec{x} - Et/\hbar)}. \tag{373}$$

Given that $\vec{k}_e = 0$, substitute (373) into (372) to yield

$$m_e c^2 \begin{pmatrix} I & \bigcirc \\ \bigcirc & I \end{pmatrix} \begin{pmatrix} \Psi_+ \\ \Psi_- \end{pmatrix} = E \begin{pmatrix} \Psi_+ \\ \Psi_- \end{pmatrix}$$

where

$$\chi_0 = \begin{pmatrix} \Psi_+ \\ \Psi_- \end{pmatrix} \qquad 2 \times 2\text{-component spinors}$$

$$\text{spinor:} \quad \left. \begin{array}{l} \Psi_+ = \psi_+^+ \alpha_+ + \psi_-^+ \alpha_- \\ \Psi_- = \psi_+^- \alpha_+ + \psi_-^- \alpha_- \end{array} \right\} \alpha_+ = \begin{pmatrix} 1 \\ 0 \end{pmatrix}, \quad \alpha_- = \begin{pmatrix} 0 \\ 1 \end{pmatrix}.$$

Thus

$$\left. \begin{array}{l} m_e c^2 \Psi_+ = E\Psi_+ \Rightarrow E = m_e c^2 \\ -m_e c^2 \Psi_- = E\Psi_- \Rightarrow E = -m_e c^2 \end{array} \right\} \Delta E = 2m_e c^2.$$

The negative rest mass energy led Dirac to propose the existence of the positron. The energy difference is the difference in rest mass energies of electrons and positrons.

13.2.3 Non-relativistic limit of Dirac's equation

By examining the behavior of the Dirac equation in the non-relativistic limit we can simplify the Hamiltonian but retain the effects of spin. Let

$$\mathcal{E} \equiv i\hbar \frac{\partial}{\partial t} - m_e c^2 \tag{374}$$

$$\vec{P} \equiv -i\hbar\vec{\nabla} + \frac{e}{c}\vec{A}. \tag{375}$$

Consider the limit of (371) when $\mathcal{E}$ and $\vec{P}$ are small compared to $m_e c^2$ and $m_e c$ respectively.

Let us define wave function χ such that

$$\chi = \begin{pmatrix} \Psi_+ \\ \Psi_- \end{pmatrix} \tag{376}$$

so that (371) becomes

$$(\mathcal{E} + e\phi)\Psi_+ - c\vec{\sigma} \cdot \vec{P}\Psi_- = 0 \tag{377}$$

$$(\mathcal{E} + 2m_ec^2 + e\phi)\Psi_- - c\vec{\sigma} \cdot \vec{P}\Psi_+ = 0 \tag{378}$$

where

$$\vec{\sigma} \equiv \sigma_x\hat{x} + \sigma_y\hat{y} + \sigma_z\hat{z}.$$

Equation (378) can be written as

$$\Psi_- = [2m_ec^2 + (\mathcal{E} + e\phi)]^{-1}c\vec{\sigma} \cdot \vec{P}\Psi_+. \tag{379}$$

For $\mathcal{E} + e\phi \ll 2m_ec^2$

$$\Rightarrow [2m_ec^2 + (\mathcal{E} + e\phi)]^{-1} \approx \frac{1}{2m_ec^2}\left(1 - \frac{\mathcal{E} + e\phi}{2m_ec^2}\right).$$

Substituting this into (377)

$$\Rightarrow \left[(\mathcal{E} + e\phi) - \frac{\vec{\sigma} \cdot \vec{P}}{2m_e}\left(1 - \frac{\mathcal{E} + e\phi}{2m_ec^2}\right)\vec{\sigma} \cdot \vec{P}\right]\Psi_+ = 0 \tag{380}$$

correct to order v^2/c^2!

To simplify further we factor out the dependence on rest energy so that

$$\mathcal{E}\Psi_+ = \mathcal{E}(\Psi(\vec{x}, t)\,e^{-im_ec^2t/\hbar}) = i\hbar\frac{\partial\Psi}{\partial t}\,e^{-im_ec^2t/\hbar}.$$

Substituting into (380) with the definition of $\mathcal{E}$ and after a little manipulation all the m_ec^2 terms cancel out so that

$$\Rightarrow \mathcal{S}\Psi = 0 \qquad \mathcal{S} = \mathcal{S}_0 + \mathcal{S}_2 \tag{381}$$

where

$$\mathcal{S}_0 = \left(i\hbar\frac{\partial}{\partial t} + e\phi\right) - \frac{(\vec{\sigma} \cdot \vec{P})^2}{2m_e}$$

and

$$\mathcal{S}_2 = \frac{\vec{\sigma} \cdot \vec{P}}{2m_ec}\left(i\hbar\frac{\partial}{\partial t} + e\phi\right)\frac{\vec{\sigma} \cdot \vec{P}}{2m_ec}.$$

In the fully non-relativistic limit $\mathcal{S}_0 \gg \mathcal{S}_2$ so that

$$\Rightarrow \mathcal{S}_0\Psi = 0. \tag{382}$$

Since σ operates in a different space than $\vec{P}$

$$(\vec{\sigma} \cdot \vec{P})^2 = \vec{P} \cdot \vec{\sigma}\vec{\sigma} \cdot \vec{P} \tag{383}$$

$$\Rightarrow \sigma_i \sigma_k = \frac{1}{2} \left[(\sigma_i \sigma_k + \sigma_k \sigma_i) + (\sigma_i \sigma_k - \sigma_k \sigma_i) \right]$$

$$= \delta_{ik} + i\epsilon_{ikm}\sigma_m$$

which implies

$$\vec{P} \cdot \vec{\sigma}\vec{\sigma} \cdot \vec{P} = |\vec{P}|^2 + i\vec{\sigma} \cdot (\vec{P} \times \vec{P}) \tag{384}$$

where

$$\vec{P} = -i\hbar\vec{\nabla} + \frac{e\vec{A}}{c} \Rightarrow \vec{P} \times \vec{P} \neq 0.$$

We are now in the position of working out two do-able cases.
Case A: Discard terms of $\approx v^2/c^2$. Radiative transitions involving spin.
Case B: Consider terms of $\approx v^2/c^2$ but let $\vec{A} = 0$.
The former allows us to describe radiative transitions involving spin (Section 13.3), the latter describes spin–orbit coupling (Section 13.4).

13.3 Radiative transitions involving spin

Recall

$$\vec{P} = \vec{p} + \frac{e\vec{A}}{c} = -i\hbar\vec{\nabla} + \frac{e\vec{A}}{c} \qquad \vec{A} \neq 0$$

so that (384) becomes

$$\vec{P} \cdot \vec{\sigma}\vec{\sigma} \cdot \vec{P} = \left| \vec{p} + \frac{e}{c}\vec{A} \right|^2 + \frac{e\hbar}{c}\vec{\sigma} \cdot (\vec{\nabla} \times \vec{A}). \tag{385}$$

So take (381) for $\mathcal{S}_0$ and put it into Hamiltonian form

$$\Rightarrow \mathcal{S}_0\Psi = 0 \qquad \Rightarrow (H_0 + H_1 + H_2 + H_{BS})\Psi = i\hbar\frac{\partial\Psi}{\partial t} \tag{386}$$

where, as before

$$H_0 + H_1 + H_2 = \frac{1}{2m_e}\left| \vec{p} + \frac{e}{c}\vec{A} \right|^2 - e\phi$$

and H_{BS} contains the spin term in (385)

$$H_{BS} = \frac{e\hbar}{2m_ec}\vec{\sigma} \cdot (\vec{\nabla} \times \vec{A}) = \frac{e\hbar}{2m_ec}\left(\frac{2}{\hbar}\vec{s} \right) \cdot \vec{B}$$

$$\Rightarrow H_{BS} = \frac{e}{m_ec}\vec{s} \cdot \vec{B}. \tag{387}$$

As we noted earlier, $\vec{s}$ acts as a spin-angular-momentum operator, so that

$$\left(s_x^2 + s_y^2 + s_z^2\right)\alpha\pm = \tfrac{3}{4}\hbar^2\alpha\pm$$

$$s_z\alpha\pm = \pm\tfrac{1}{2}\hbar\alpha\pm$$

$$(s_x \pm i s_y)\alpha\mp = \hbar\alpha\pm$$

$$(s_x \pm i s_y)\alpha\pm = 0.$$

Define $\pm\tfrac{1}{2}$ terms with $m_s = \tfrac{1}{2}$ (spin up), $= -\tfrac{1}{2}$ (spin down). Now suppose

$$\vec{A} = \vec{A}_{rad} + \frac{1}{2}\vec{B}_0 \times \vec{x}. \tag{388}$$

The modified potential still satisfies the Coulomb gauge but $\vec{\nabla} \times \vec{A} \rightarrow$ static field

$$\Rightarrow H_{BS} = H_{BS}^{rad} + H_{zs}$$

$$H_{BS}^{rad} = (e/m_e c)\vec{s} \cdot (\vec{\nabla} \times \vec{A}_{rad}) \tag{389}$$

$$H_{zS} = (e/m_e c)\vec{s} \cdot \vec{B}_0.$$

The H_{BS}^{rad} does not contribute to the transition matrix element in the same multipole order as the largest term present in H. Note the parallel between $\hat{k} \times \hat{e}_\alpha$ and $\vec{\nabla} \times \vec{A}_{rad} = \vec{B}_{rad}$. We see that magnetic dipoles interact only with the $\vec{B}$ part of the EM field.

13.3.1 *Zeeman effect*

The previous discussion allows us to describe the Zeeman effect. We begin by examining the static B_0 portion of H_1, and express the Hamiltonian as

$$H_{zL} = \frac{e}{2m_e c}(\vec{B}_0 \times \vec{x}) \cdot \vec{p} = \frac{e}{2m_e c}\vec{B}_0 \cdot \vec{L}.$$

Then, we add H_{zS} to get the total H_z,

$$H_z = \frac{-e B_0}{2m_e c} \cdot (\vec{L} + \vec{s}) = -\vec{B}_0 \cdot (\vec{M} + \vec{m}). \tag{390}$$

The Zeeman effect is an example of spin coupling.

13.4 **Relativistic correction with $A = 0$**

Letting $A = 0$ reduces $\vec{P}$ to the canonical momentum, $\vec{p}$, so that

$$\vec{P} = \vec{p} \Rightarrow \vec{p} \cdot \vec{\sigma}\vec{\sigma} \cdot \vec{p} = |\vec{p}|^2.$$

Retaining the relativistic corrections to v^2/c^2 yields

$$S = S_0 + S_2 \rightarrow S' = S_0 + 1/(8m_e^2 c^2)\{e\hbar(\nabla^2\phi) + |\vec{p}|^4/m_e$$
$$+ 2e\hbar\vec{\sigma} \cdot [(\vec{\nabla}\phi) \times (\vec{p})]\} \tag{391}$$
$$\Rightarrow \text{solve} \qquad S'\Psi' = 0.$$

In Hamiltonian form $S' = H - i\hbar(\partial/\partial t)$, so if we compare with (391)

$$\Rightarrow H_S = S_0 - i\hbar\frac{\partial}{\partial t} = i\hbar\frac{\partial}{\partial t} + e\phi - \frac{|\vec{p}|^2}{2m_e} - i\hbar\frac{\partial}{\partial t}$$

we get

$$\Rightarrow H_S = \frac{|\vec{p}|^2}{2m_e} - e\phi \tag{392}$$

the structure Hamiltonian

$$H_k = \frac{|\vec{p}|^4}{8m_e^2 c^2} \tag{393}$$

the relativistic correction for kinetic energy

$$H_0 = \frac{-e\hbar^2}{8m_e^2 c^2}\nabla^2\phi \tag{394}$$

the relativistic correction for the potential, affecting only the $\ell = 0$ term and

$$H_{SO} = \frac{-e\vec{s} \cdot [(\vec{\nabla}\phi) \times \vec{p}]}{2m_e c^2} \tag{395}$$

the term representing spin–orbit coupling where $\vec{s} = (\hbar/2)\vec{\sigma}$. Note that spin–orbit coupling is like a self induced Zeeman splitting because it is the motion of the electron in the potential that produces a local B field (recall Ampere's law). Spin–orbit coupling accounts for the *fine structure of atoms*.

13.5 Atomic fine structure

I now examine the nature of the spin–orbit interaction. In the process, I will solve for the energy eigenvalues of the spin–orbit Hamiltonian. I conclude by describing the fine structure of the atomic energy levels of hydrogen.

13.5.1 Spin–orbit interaction

The total angular momentum is now the sum of spin and orbital angular momenta

$$\vec{J} = \vec{L} + \vec{S}$$

which requires a change of counters: $m \to m_\ell$, and addition of m_s, j so that the new eigenket is $|n, \ell, j, m_\ell, m_s\rangle$.

13.5.2 Time-independent perturbation theory

Spin lifts the degeneracy with respect to ℓ and m_ℓ.

We want to solve

$$(H_0 + H_{SO})\,\Phi = E\Phi$$

where, from (395)

$$H_{SO} \propto \vec{S} \cdot (\vec{\nabla}\phi \times \vec{p}) \propto \vec{S} \cdot \left(\frac{1}{r}\frac{dV}{dr}\vec{r} \times \vec{p} \right) \propto \frac{1}{r}\frac{dV}{dr}\vec{S} \cdot \vec{L}$$

so that

$$H_{SO} = \frac{1}{2m_e^2 c^2} \left(\frac{1}{r}\frac{dV}{dr} \right) \vec{S} \cdot \vec{L}. \tag{396}$$

Let Φ be a linear superposition of unperturbed states

$$\Phi = \sum_{m'_\ell, m'_s} C(m'_\ell, m'_s) |n, \ell, m'_\ell, m'_s\rangle$$

where

$$|n, \ell, m'_\ell, m'_s\rangle = \frac{1}{r} R_{n,\ell}(r) Y_{\ell, m'_\ell} \alpha_\pm.$$

Substituting the above into (396) yields

$$\sum_{m'_\ell, m'_s} (E_{n\ell} - E + H_{SO})|n, \ell, m'_\ell, m'_s\rangle C(m'_\ell, m'_s) = 0.$$

Left multiply by $\langle n, \ell, m_\ell, m_s|$

$$\sum_{m'_\ell, m'_s} (-E_{SO}\delta_{m_\ell,m'_\ell}\delta_{m_s,m'_s} + \langle n, \ell, m_\ell, m_s|H_{SO}|n, \ell, m'_\ell, m'_s\rangle) C(m'_\ell, m'_s) = 0$$

where

$$E_{SO} \equiv E - E_0 = E - E_{n\ell}.$$

Since $\alpha_\pm$ is a two-column matrix we can cast the above as

$$\begin{pmatrix} A_{++} & A_{+-} \\ A_{-+} & A_{--} \end{pmatrix} \begin{pmatrix} C_+ \\ C_- \end{pmatrix} = 0. \tag{397}$$

Each A has $(2\ell + 1)^2$ matrix elements from the $2\ell + 1$ values of m_l in C. Thus

$$A_{++} = \begin{bmatrix} H_{\ell\ell}^{++} - E_{SO} & H_{\ell,\ell-1}^{++} & \cdots & H_{\ell,-\ell}^{++} \\ H_{\ell-1,\ell}^{++} & H_{\ell-1,\ell-1}^{++} - E_{SO} & \cdots & H_{\ell-1,-\ell}^{++} \\ \cdots & \cdots & \cdots & \cdots \\ H_{-\ell,\ell}^{++} & H_{-\ell,\ell-1}^{++} & \cdots & H_{-\ell,-\ell}^{++} - E_{SO} \end{bmatrix} \tag{398}$$

where

$$H_{m_\ell m'_\ell}^{\pm\pm} \equiv \langle n, \ell, m_\ell, \pm 1/2 | H_{SO} | n, \ell, m'_\ell, \pm 1/2 \rangle \tag{399}$$

and

$$C_+ \equiv \begin{bmatrix} C(\ell, +1/2) \\ C(\ell-1, +1/2) \\ \cdot \\ \cdot \\ \cdot \\ C(-\ell, +1/2) \end{bmatrix} \qquad C_- \equiv \begin{bmatrix} C(\ell, -1/2) \\ C(\ell-1, -1/2) \\ \cdot \\ \cdot \\ \cdot \\ C(-\ell, -1/2) \end{bmatrix}. \tag{400}$$

The requirement for solving (397) is

$$\begin{vmatrix} A_{++} & A_{+-} \\ A_{-+} & A_{--} \end{vmatrix} = 0 \tag{401}$$

which yields a characteristic equation for E_{SO} as a polynomial of order $2(2\ell + 1)$. The roots are eigenvalues corresponding to the perturbation energies.

13.5.3 *The jm representation*

Substituting (396) into (399) yields

$$H_{m_\ell m'_\ell}^{\pm\pm} = \left[\frac{1}{2m_e^2 c^2} \int_0^\infty \frac{1}{r} \frac{dV}{dr} R_{n\ell}^2(r)\, dr \right] \langle Y_{\ell m_\ell} \alpha(m_s) | \vec{S} \cdot \vec{L} | Y_{\ell m'_\ell} \alpha(m'_s) \rangle \tag{402}$$

where

$$\vec{S} \cdot \vec{L} \equiv S_x L_x + S_y L_y + S_z L_z.$$

Note that $S_x L_x + S_y L_y = \frac{1}{2}[(S_x + iS_y)(L_x - iL_y) + (S_x - iS_y)(L_x + iL_y)]$.
These are raising and lowering operators so that

$$(L_x \pm iL_y)Y_{lm} = [(l \mp m)(\ell \pm 1 + m)]^{1/2} \hbar Y_{\ell, m\pm 1}$$

and

$$(S_x \pm i S_y)\alpha_{\mp} = \hbar \alpha_{\pm}$$

so that

$$\langle Y_{\ell,m_\ell}\alpha(m_s)|S_x L_x + S_y L_y|Y_{\ell,m'_\ell}\alpha(m'_s)\rangle = 0$$

unless $m'_s + m'_\ell = m_s + m_\ell$, while

$$\langle Y_{\ell,m_\ell}\alpha(m_s)|S_z L_z|Y_{\ell,m'_\ell}\alpha(m_s)\rangle = 0$$

unless $m'_s = m_s$ and $m'_\ell = m_\ell$, consistent with the above constraint. Thus

$$H^{\pm\pm}_{m_\ell m'_\ell} = \begin{cases} 0, & m'_s + m'_\ell \neq m_s + m_\ell \\ \neq 0, & m'_s + m'_\ell = m_s + m_\ell = m \end{cases}$$

$\Rightarrow m = m_s + m_\ell$ is a *good* quantum number.

Since $J^2\Phi = j(j+1)\hbar^2\Phi$, j is also a good quantum number.

Given that $J^2 = S^2 + L^2 + 2\vec{S}\cdot\vec{L}$ we have

$$[j(j+1)]_{\max} = s(s+1) + \ell(\ell+1) + 2\ell s$$
$$[j(j+1)]_{\min} = s(s+1) + \ell(\ell+1) - 2\ell s$$

so that $\ell - s \leq j \leq \ell + s$ in increments of 1!

Thus, *in the jm representation*, the new eigenket becomes $|n, \ell, j, m\rangle$.

13.5.4 Solution for E_{SO}

We can now drop the $n\ell$ from the eigenket so that $\Phi = |jm\rangle$. Substituting into the Schrödinger equation yields

$$E = \langle jm|H_0|jm\rangle + \langle jm|H_{SO}|jm\rangle = E_0 + E_{SO}$$

where

$$E_{SO} = \left[\frac{1}{2m_e^2 c^2}\int_0^\infty \frac{1}{r}\frac{dV}{dr}R_{n\ell}^2\, dr\right]\langle jm|\vec{S}\cdot\vec{L}|jm\rangle.$$

But $\vec{S}\cdot\vec{L} = 1/2[J^2 - L^2 - S^2]$ is diagonal in the *jm* representation

$$\Rightarrow \langle jm|\vec{S}\cdot\vec{L}|jm\rangle = \frac{1}{2}[j(j+1) - \ell(\ell+1) - s(s+1)]\hbar^2$$

$$= \begin{cases} (\ell/2)\hbar^2, & j = \ell + \frac{1}{2} \\ -(\ell+1)\hbar^2/2, & j = \ell - \frac{1}{2}. \end{cases}$$

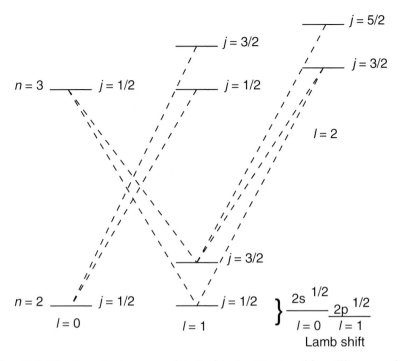

Fig. 13.1 The fine structure associated with the $H\alpha$ transition. The expanded region shows the Lamb Shift which is not predicted by Dirac theory.

For a hydrogen atom, $V = -e^2/r$. Substituting into the previous equation and grouping terms

$$\Rightarrow E_{SO} = |E_n| \frac{\alpha^2}{n\ell\left(\ell + \frac{1}{2}\right)(\ell + 1)} \begin{cases} \ell/2, & j = \ell + \frac{1}{2} \\ -(\ell + 1)/2, & j = \ell - \frac{1}{2}. \end{cases}$$

The corresponding perturbation from the relativistic correction is

$$E_k = -|E_n| \frac{\alpha^2}{n^2} \left(\frac{n}{\ell + \frac{1}{2}} - \frac{3}{4} \right).$$

Adding E_{SO} and E_k yields

$$\Delta E = E_k + E_{SO} = -|E_n| \frac{\alpha^2}{n^2} \left(\frac{n}{j + \frac{1}{2}} - \frac{3}{4} \right). \tag{403}$$

For $j = 1/2, 3/2, \ldots, n - (1/2)$.

We see that levels are still degenerate with respect to m but not with respect to j. Applying (403) to the $n = 4$ level of hydrogen, as an example, we see that there are now four levels

$$\Delta E = -|E_4|\alpha^2 \begin{cases} \frac{13}{16}, & j = \frac{1}{2} \\[2mm] \frac{5}{64}, & j = \frac{3}{2} \\[2mm] \frac{7}{176}, & j = \frac{5}{2} \\[2mm] \frac{1}{64}, & j = \frac{7}{2}. \end{cases}$$

Spin–orbit coupling produces fine structure of the order of α^2 of the unperturbed energy. The new dipole selection rules are

$$\Delta m = 0, \pm 1 \qquad \Delta \ell = \pm 1 \qquad \Delta j = 0, \pm 1.$$

The fine structure associated with the $H\alpha$ transition is shown in Fig. 13.1.

13.6   Further reading

Cohen-Tanoudji, C. (1992) *Principles of Quantum Mechanics*, John Wiley and Sons, New York, NY, USA.

Dirac, P.A.M. (1982) *Principles of Quantum Mechanics*, 4th edn, Oxford University Press, Oxford, UK.

Griffiths, D. (1995) *Introduction to Quantum Mechanics*, Prentice Hall, Englewood Cliffs, NJ, USA.

Landau, L. D. and Lifschitz, E. M. (1977) *Quantum Mechanics*, 3rd edn, Pergamon Press, Oxford, UK.

Landau, R. H. (1996) *Quantum Mechanics II: A Second Course in Quantum Theory*, 2nd edn, John Wiley and Sons, New York, NY, USA.

Osterbrock, D. E. (1974) *Astrophysics of Gaseous Nebulae*, W. H. Freeman, San Francisco, CA, USA.

Shu, F. H. (1992) *The Physics of Astrophysics: Radiation*, University Science Books, Mill Valley, CA, USA.

Chapter 14

Atomic hyperfine lines

In expanding the relativistic Hamiltonian we noted two do-able cases. In the first case, we demonstrated that interactions of spin and orbital momentum with magnetic fields can take place thereby producing fine structure. In the second case, we showed that an electron's spin can interact with its orbital momentum (self induced Zeeman effect) to produce atomic fine structure. Of special interest in Case A is the spin–spin interaction. In that case the electron's spin interacts with the magnetic field produced by the spin of the nucleus. The spin–spin interaction is therefore an example of the Zeeman effect which leads to hyperfine structure. We now proceed to examine the spin–spin interaction in the ground state $(1S_{1/2})$ of hydrogen. Note that spin–orbit coupling alone does not split this state.

14.1 The 21 cm line of hydrogen

The spin magnetic moments of the electron and nucleus are given by

$$\vec{m} = -\frac{g_e e}{2m_e c}\vec{s} \tag{404}$$

$$\vec{M} = \frac{g_p e}{2m_p c}\vec{S} \tag{405}$$

respectively, where g_e and g_p are the so-called g factors of the electron and nucleus. Associated with $\vec{M}$ is a magnetic field. Recall that for a dipole field

$$\vec{B} = \frac{3(\vec{M} \cdot \hat{k})\hat{k} - \vec{M}}{|\vec{x}|^3} \tag{406}$$

where $\hat{k}$ represents the direction of the dipole moment and $\vec{x}$ is the distance from the center of the dipole.

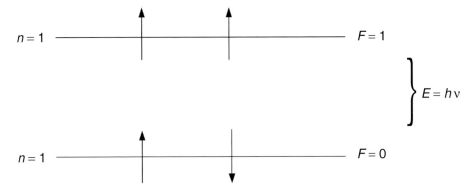

Fig. 14.1 Hyperfine splitting of the ground state of hydrogen. Transitions between the $F = 1$ and $F = 0$ states lead to the 21 cm line. Shown are the relative alignments of the proton and electron spins.

According to our earlier discussions we can calculate the Zeeman effect

$$H_m = H_{zS} = \frac{e}{m_e c} \vec{s} \cdot \vec{B} = -\vec{m} \cdot \vec{B}. \tag{407}$$

Since the proton (nucleus) and electron each have a spin of $1/2$ there are four possible combinations for the combined wave function. We define a total spin $\vec{F}$ such that $\vec{F} = \vec{S} + \vec{s}$. Therefore there are three possible states that produce $F = 1$ and one state that produces $F = 0$ (Fig. 14.1)

$$\Sigma_{F_z}^{F^2} = \Sigma_{\pm 1}^1, \Sigma_0^1, \Sigma_0^0 \tag{408}$$

where

$$\Sigma_{\pm 1}^1 = \alpha_{\pm}(p)\alpha_{\pm}(e)$$

$$\Sigma_0^1 = \frac{1}{\sqrt{2}}[\alpha_+(p)\alpha_-(e) + \alpha_-(p)\alpha_+(e)] \tag{409}$$

$$\Sigma_0^0 = \frac{1}{\sqrt{2}}[\alpha_+(p)\alpha_-(e) - \alpha_-(p)\alpha_+(e)].$$

The energies of the levels can be calculated in the usual way by taking the expectation value of the Hamiltonian

$$E_m = \langle \phi \Sigma | H_m | \phi \Sigma \rangle \tag{410}$$

where ϕ is the unperturbed wave function for the ground state of hydrogen and is given by

$$\phi = 2r_0^{-3/2} e^{-r/r_0} Y_{00} \tag{411}$$

while Σ corresponds to the four spin states previously described.

Direct substitution of equations (404) and (411) into (410) results in a divergent integral. We therefore have to resort to a trick first formulated by Fermi.

Fermi's trick

We know that for a magnetic dipole (see Jackson, 1998)

$$\vec{A} = \vec{\nabla} \times \left(\frac{\vec{M}}{r} \right) \tag{412}$$

so that the B field can be written as

$$\vec{B} = \vec{\nabla} \times \vec{A} \rightarrow \nabla \times \left[\vec{M} \times \vec{\nabla} \left(\frac{1}{r} \right) \right]. \tag{413}$$

From (407), $H_m = -\vec{m} \cdot \vec{B}$. Inserting (413) yields

$$H_m = (\vec{m} \cdot \vec{M})\nabla^2 \left(\frac{1}{r} \right) - [(\vec{m} \cdot \vec{\nabla})(\vec{M} \cdot \vec{\nabla})]\frac{1}{r}.$$

Subtracting one-third of the first term from the first term and adding this to the second term yields

$$H_m = H_m^{(0)} + H_m^{(2)}$$

where

$$H_m^{(0)} = \frac{2}{3}(\vec{m} \cdot \vec{M})\nabla^2 \left(\frac{1}{r} \right) \tag{414}$$

and

$$H_m^{(2)} = -[(\vec{m} \cdot \vec{\nabla})(\vec{M} \cdot \nabla) - 1/3(\vec{m} \cdot \vec{M})\nabla^2]\frac{1}{r}.$$

We see that the bad behavior at $r = 0$ is localized to $H_m^{(2)}$. However, when inserted into (410) it yields only spherical harmonics, but because it is operating on a spherically symmetric function (411) the net result from the operation is 0. Thus

$$\langle \phi \Sigma | H_m^{(2)} | \phi \Sigma \rangle = 0$$

and takes away the bad behavior at $r = 0$. Thus the remaining expression is

$$E_m = \langle \phi \Sigma | H_m^{(0)} | \phi \Sigma \rangle.$$

Using $\nabla^2(1/r) = -4\pi\delta(\vec{x})$ with (404) and (414), this reduces to

$$E_m = \frac{2 g_e g_p e^2}{3 m_e m_p c^2 a_0^3} \langle \Sigma | \vec{S} \cdot \vec{s} | \Sigma \rangle = \frac{2}{3} \frac{E_0}{\hbar^2} \langle \Sigma | \vec{S} \cdot \vec{s} | \Sigma \rangle. \tag{415}$$

Now, $\vec{S} \cdot \vec{s} = \frac{1}{2}(\vec{F} \cdot \vec{F} - \vec{S} \cdot \vec{S} - \vec{s} \cdot \vec{s})$ so that

$$\langle \Sigma | \vec{S} \cdot \vec{s} | \Sigma \rangle = \frac{1}{2}[\langle \Sigma | \vec{F} \cdot \vec{F} | \Sigma \rangle - \langle \Sigma | \vec{S} \cdot \vec{S} | \Sigma \rangle - \langle \Sigma | \vec{s} \cdot \vec{s} | \Sigma \rangle]$$

$$= \frac{1}{2}[F(F+1)\hbar^2 - S(S+1)\hbar^2 - s(s+1)\hbar^2].$$

For $F = 1$, $E_{m1} = (1/6)E_0$ and for $F = 0$, $E_{m2} = -(1/2)E_0$, where E_0 can be rewritten as

$$E_0 = g_e g_p \alpha^4 \left(\frac{m_e}{m_p} \right) m_e c^2.$$

The net energy difference between the $F = 1$ and $F = 0$ states is given by

$$\Delta E = E_{m1} - E_{m2} = \frac{2}{3} E_0 = 6 \times 10^{-6} \text{eV}.$$

For $h\nu = \Delta E \rightarrow \nu = 1.420\,405\,752 \times 10^9$ Hz $\rightarrow \lambda \approx 21$ cm.

14.1.1 Transition rate

Recall that it is the total angular momentum that contributes to the Zeeman effect. In this case the total momentum is $S + s$. Thus, in analogy to the electric dipole transition, we can define a magnetic dipole transition as

$$A_{if} = \frac{4\omega^3}{3\hbar c^3} |\langle f | \vec{M} + \vec{m} | i \rangle|^2 \tag{416}$$

which we get by setting $e\vec{x} \rightarrow \vec{M} + \vec{m}$. Now recall that the spin can be expressed in terms of the Pauli spin matrices. Thus

$$\vec{M} = \mu_p \vec{\sigma}_p \quad \text{and} \quad \vec{m} = -\mu_e \vec{\sigma}_e$$

where

$$\mu_p = \frac{g_p e \hbar}{4 m_p c} \quad \text{and} \quad \mu_e = -\frac{g_e e \hbar}{4 m_e c}.$$

For the $F = 1 \rightarrow 0$ transition

$$A_{10} = \frac{4\omega^3}{3\hbar c^3} \left| \left\langle \Sigma_0^0 | \mu_p \vec{\sigma}_p - \mu_e \vec{\sigma}_e | \Sigma^1 \right\rangle \right|^2. \tag{417}$$

Recall that $\sigma = \sigma_x \hat{x} + \sigma_y \hat{y} + \sigma_z \hat{z}$, where $\sigma_z \alpha_\pm = \pm \alpha_\pm$ and $\sigma_x \alpha_\pm = \alpha_\pm$, $\sigma_y \alpha_\pm = \pm i \alpha_\mp$ so that

$$\left| \left\langle \Sigma_0^0 | \mu_p \vec{\sigma}_p - \mu_e \vec{\sigma}_e | \Sigma_{\pm 1}^1 \right\rangle \right| = -\frac{1}{\sqrt{2}} (\hat{x} \pm i \hat{y})(\mu_p + \mu_e)$$

$$\left| \left\langle \Sigma_0^0 | \mu_p \vec{\sigma}_p - \mu_e \vec{\sigma}_e | \Sigma_0^1 \right\rangle \right| = \hat{z}(\mu_p + \mu_e).$$

Table 14.1. *Transitions involving hyperfine structure*

Atom	Spin	Transition	Frequency (Hz)	$A\,(\mathrm{s}^{-1})$
HI	1/2	$^2S_{1/2},\ F = 0 - 1$	1.420405751×10^9	2.85×10^{-15}
D	1	$^2S_{1/2},\ F = 1/2 - 3/2$	3.27384349×10^8	4.64×10^{-17}
HeII	1/2	$^2S_{1/2},\ F = 1 - 0$	8.66566×10^9	6.50×10^{-13}
NVII	1	$^2S_{1/2},\ F = 1/2 - 3/2$	5.306×10^7	1.49×10^{-19}
NI	1	$^4S_{3/2},\ F = 3/2 - 5/2$	2.612×10^7	1.78×10^{-20}
		$F = 1/2 - 3/2$	1.567×10^7	3.84×10^{-21}
H_2^+	1	$F_2,\ F = 3/2, 5/2 - 1/2, 3/2$	1.40430×10^9	2.75×10^{-15}
		$F_2,\ F = 3/2, 3/2 - 1/2, 3/2$	1.41224×10^9	2.80×10^{-15}
NaI	3/2	$^2S_{3/2},\ F = 1 - 2$	1.77161×10^9	5.56×10^{-15}

Recalling that the transition rate is independent of the upper level state we choose (when states are degenerate), we have

$$A_{10} = \frac{4\omega^3}{3\hbar c^3}\left|\langle \Sigma_0^0 | \mu_p \vec{\sigma}_p - \mu_e \vec{\sigma}_e | \Sigma^1 \rangle\right|^2 = \frac{4\omega^3}{3\hbar c^3}(\mu_p + \mu_e)^2. \qquad (418)$$

But $\mu_e \gg \mu_p$ so that

$$A_{10} = \frac{g_e}{12}\left(1 + \frac{g_p m_e}{g_e m_p}\right)^2 \left(\frac{r_e\omega}{c}\right)\left(\frac{\hbar\omega}{m_e c^2}\right)\omega \qquad (419)$$

$$\Rightarrow A_{10} \approx 3 \times 10^{-15}\mathrm{s}^{-1}.$$

The lifetime of the excited state is $\approx 10^7$ years.

Note that the hyperfine structure scales as α^4 while the fine structure scales as α^2. The 21 cm line represents the most famous astrophysical transition arising from hyperfine structure. Other hyperfine transitions are also possible and these are summarized in Table 14.1.

In direct analogy to (384) for the fine structure, the general formula for calculating energy hyperfine splitting is given by

$$E_{SS} = g_N(m_e m_p)\frac{\alpha^2 hcRZ^3}{n^3}\left[\frac{F(F+1) - S(S+1) - j(j+1)}{j(j+1)(2j+1)}\right]. \qquad (420)$$

14.1.2 The 21 cm line profile

The natural width of the 21 cm line is $\Gamma \approx A_{10} = 3 \times 10^{-15}\ \mathrm{s}^{-1}$. The extremely narrow width follows from the uncertainty principle because of the extremely long lifetime of the excited state. The lifetime is $\approx A_{10}^{-1} = 10^7$ years. The naturally narrow width of this line makes it an ideal diagnostic of interstellar hydrogen because the strength of the line depends on the density and temperature of the

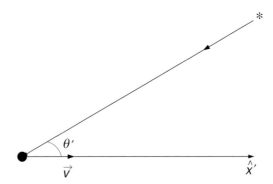

Fig. 14.2 The geometry associated with the Doppler effect.

hydrogen gas. The latter broadens the line because of differential Doppler shifts from randomly moving hydrogen atoms. In the next section we review the Doppler effect and then apply it to determine the profile of the 21 cm line.

14.2   The Doppler effect

Consider the motion of an observer relative to a star, as shown in Fig. 14.2. Recall from special relativity that energy, E, and momentum, $\vec{p}$, make up a four-vector. The E is the fourth component of such a four-vector and therefore transforms as

$$E = \gamma(V)[E' + Vp'_x]$$

in analogy to

$$t = \gamma(V)\left[t' + \frac{V}{c^2}x'\right]$$

where the primed coordinates refer to the frame of the star. Thus

$$\Rightarrow E = \gamma(V)\left[E' + \frac{E'}{c}V_{\parallel}\right]$$

where the radial velocity with respect to the star is defined as

$$V_{\parallel} = V \cos\theta'$$

so that

$$E = \gamma(V)E'\left[1 + \frac{V}{c}\cos\theta'\right].$$

For photons we can let $E = h\nu$, $E' = h\nu'$

$$\nu = \nu'\gamma(V)[1 + \beta\cos\theta'].$$

All that remains is to transform $\theta' \to \theta$, which we can do by using the *aberration* formulae, so that

$$= v'\gamma \left[1 + \beta \left(\frac{\cos\theta - \beta}{1 - \beta\cos\theta} \right) \right].$$

Now, let us consider some special cases

$$\theta = 0 \; : \; v = v'\gamma(1 + \beta) = v' \left[(1 + \beta)/(1 - \beta) \right]^{1/2}$$

$$\theta = \pi/2 : v = v'\gamma(1 - \beta^2) = v'[1 - \beta^2]^{1/2}.$$

We see that at relativistic velocities there is a significant *transverse* Doppler shift in addition to the regular radial shift. The transverse shift is always a redshift. In the non-relativistic limit, $\beta \ll 1$

$$v = v' \begin{cases} 1 + \beta, & \theta = 0 \\ 1 - \frac{1}{2}\beta^2, & \theta = \pi/2. \end{cases} \tag{421}$$

Thus, to first order in β there is only a radial Doppler shift.

14.2.1 Doppler broadening of the 21 cm line

The 21 cm line emissivity of HI gas is given by

$$j_v = \frac{hv}{4\pi} n_1 A_{10} \phi_v$$

where n_1 is the number density of HI in the $F = 1$ state and ϕ_v is the normalized line profile such that

$$\int_0^\infty \phi_v \, dv = 1.$$

Since HI atoms are in collisional equilibrium with each other they follow a Maxwell–Boltzmann velocity distribution. The probability, therefore, of an atom's radial velocity being between v and $v + dv$ at a temperature T is given by

$$4\pi f(v)v^2 \, dv = 4\pi \left(\frac{m_H}{2\pi kT} \right)^{1/2} e^{-(m_H v^2)/(2kT)} v^2 \, dv \tag{422}$$

where

$$4\pi \int_0^\infty f(v)v^2 \, dv = 1.$$

At interstellar temperatures the HI gas is not relativistic so that (421) can be used for the Doppler shift, in which case

$$\frac{\Delta v}{v_0} = \frac{v}{c}.$$

Combining with (422)

$$f(v) = \left(\frac{m_H}{2\pi kT}\right)^{3/2} e^{-(m_H c^2 \Delta v^2)/(2kT v_0^2)}. \tag{423}$$

Similarly

$$4\pi v^2 \, dv \rightarrow 4\pi \left[\frac{\Delta v c^2}{v_0^2}\right] \frac{c}{v_0} \, dv = 4\pi \left[\frac{2kT}{m_H}\right] \frac{c}{v_0} \, dv.$$

Combining the last two equations, and insuring normalization, finally yields

$$\phi_v = \left[\frac{m_H c^2}{2\pi kT v_0^2}\right]^{1/2} e^{-(m_H c^2 (v - v_0)^2)/(2kT v_0^2)}. \tag{424}$$

This is a Gaussian whose $1/e$ halfwidth is given by

$$\sigma_v = \sqrt{\frac{2kT}{m_H}} \frac{v_0}{c}. \tag{425}$$

For $T = 100$ K, $\sigma_v = 7$ kHz for the 21 cm line.

14.3   Neutral hydrogen in galaxies

Observations of 21 cm emission can be used to probe neutral hydrogen in galaxies. I begin with a discussion of the equation of transfer for HI emission and its use in determining gas distributions in galaxies. The shape of the 21 cm line provides important information on the dynamics of the detected gas. We will therefore discuss how gas dynamics are measured. In the process, we will highlight the problem of missing mass and dark matter.

14.3.1   Equation of transfer for HI emission

Earlier, we introduced the equation of transfer. If we now define the volume absorption coefficient as

$$\kappa_v = n_H \sigma_v \, \text{cm}^{-1} \tag{426}$$

we can express the equation of transfer as

$$dI_v = j_v \, dr - \kappa_v I_v \, dr. \tag{427}$$

Recall that the optical depth is defined as $d\tau_v = \kappa_v \, dr$, so that

$$\frac{dI_v}{d\tau_v} = -I_v + \frac{j_v}{\kappa_v} \tag{428}$$

which has the solution

$$I_\nu = I_\nu(0)e^{-\tau_\nu} + e^{-\tau_\nu} \int_0^{\tau_\nu} \frac{j_\nu}{\kappa_\nu} e^{\tau_\nu'} d\tau_\nu'.$$ (429)

For the 21 cm line

$$j_\nu = \frac{h\nu}{4\pi} n_1 A_{10}\phi_\nu \text{ erg cm}^{-3} \text{ s}^{-1} \text{ sr}^{-1} \text{ Hz}^{-1}.$$ (430)

Assuming thermodynamic equilibrium

$$j_\nu = \kappa_\nu B_\nu.$$ (431)

Recall that the relationship between B and A is given by

$$B_{10} = \frac{c^2}{2h\nu^3} A_{10}.$$ (432)

Combining (430) and (432) we get

$$j_\nu = \frac{h\nu}{4\pi} n_1 \frac{2h\nu^3}{c^2} \phi_\nu B_{10}.$$ (433)

Combining (431) and (433) and recognizing that $B_{01} = 3B_{10}$

$$\kappa_\nu = \frac{j_\nu}{B_\nu} = \frac{h\nu}{4\pi} n_1 \phi_\nu (e^{h\nu/kT} - 1)B_{10}.$$ (434)

Now, from the Boltzmann relation

$$\frac{n_1}{n_0} = 3 \, e^{-h\nu/kT}.$$ (435)

Combining (434) and (435)

$$\Rightarrow \kappa_\nu = \frac{h\nu}{4\pi} \phi_\nu n_0 B_{01} \left(1 - \frac{n_1}{3n_0}\right).$$ (436)

Now let us look at the extreme cases of high and low optical depth.

Case A $\quad \tau_\nu \gg 1$

$$I_\nu \approx \frac{j_\nu}{\kappa_\nu} = \frac{2h\nu^3/c^2}{3n_0/n_1 - 1}.$$ (437)

We can now define the *spin* temperature by setting $I_\nu = B_\nu$ which leads to

$$n_1/n_0 = 3 \, e^{-h\nu/kT_s}.$$ (438)

Typically, $T_s \approx 100$ K.

At 21 cm, $h\nu \ll kT_s$ so that

$$n_0/n_1 \approx 1/3 \left(1 + \frac{h\nu}{kT_s} \right). \tag{439}$$

Combining with (437) and (438) we get

$$I_\nu = \frac{2kT_s\nu^2}{c^2}. \tag{440}$$

Equation (440) is known as the *Rayleigh–Jeans Approximation*.

Case B $\tau_\nu \ll 1$

If we set $I_\nu(0) = 0$ and integrate over all ν we get

$$I = \int I_\nu \, d\nu = \frac{h\nu}{4\pi} A_{10} \int n_1 \, dr = \frac{3h\nu}{16\pi} A_{10} \int n_H \, dr. \tag{441}$$

Use of this equation in interpreting observations of HI emission indicates that the average density of HI, $n_H \approx 0.5 \, \text{cm}^{-3}$.

14.3.2   *Emission or absorption?*

If we assume $j_\nu = \kappa_\nu B_\nu$ and T_s is constant throughout the observed region, then the equation of transfer becomes

$$I_\nu = I_\nu(0)\,e^{-\tau_\nu} + B_\nu(1 - e^{-\tau_\nu}). \tag{442}$$

Inserting the Rayleigh–Jeans Approximation (440), this becomes

$$T_b = T\,e^{-\tau_\nu} + T_s(1 - e^{-\tau_\nu}). \tag{443}$$

Typically, T represents a background source and T_s represents the foreground HI. An emission line is observed when $T_b - T > 0$, which occurs when $T < T_s$. A line is seen in absorption when $T_b - T < 0$ requiring that $T > T_s$, as you would expect from Kirchhoff's laws.

14.4   Measuring HI in external galaxies

The HI traces the building material from which stars and planets are made. External galaxies provide the optimal perspective on the distribution and kinematics of this material.

14.4.1 *Integral properties of galaxies*

The hydrogen mass of a galaxy can be estimated by integrating under the observed HI line profile. From (441) and (443) and $\tau_v \ll 1$ the surface density of HI atoms radiating in a 1 km s^{-1} Doppler interval is given by

$$N_v = 1.82 \times 10^{18} T_s \tau_v. \tag{444}$$

Since $T_b \approx T_s \tau_v$ for an optically thin gas

$$N_H = 1.82 \times 10^{18} \int_V T_b \, dV \tag{445}$$

which is the column density of hydrogen obtained by integrating over the line profile. Integration over the surface, S, of the galaxy yields the total number of HI atoms

$$\mathcal{N}_H = \int_S N_H \, dS = 1.82 \times 10^{18} \int T_b \, dV \int dS$$

$$\Rightarrow \mathcal{N}_H = 1.82 \times 10^{18} d^2 \int \int T_b \, dV \, d\Omega$$

where d is the distance to the galaxy and $dS = d^2 \, d\Omega$. In more convenient units

$$\frac{M}{M_\odot} = 2.36 \times 10^5 \left(\frac{d}{\text{Mpc}}\right)^2 \int \frac{F_v}{\text{Jy}} \frac{dV}{\text{km s}^{-1}}. \tag{446}$$

The total gravitating mass of a galaxy can be calculated by treating the HI as test particles and assuming that the galaxy is in rotational equilibrium so that

$$\frac{v^2(r)}{r} = \frac{GM(r)}{r^2} \rightarrow M(r) = \frac{rv^2(r)}{G} \tag{447}$$

$$\Rightarrow M_{\text{total}} = \frac{rv_{\text{max}}^2}{G}.$$

14.4.2 *Kinematics of the HI*

Consider a circular disk defined by the coordinates r, θ which is inclined with respect to the plane of the sky by an angle i. The observed line-of-sight velocity $u(r, \theta)$ is given by

$$u(r, \theta) = v_0 + v(r, \theta) \sin i \cos \theta + w(r, \theta) \sin i \sin \theta + z(r, \theta) \cos i \tag{448}$$

where, v_0 is the systemic velocity, $v(r, \theta)$ is the tangential velocity in the plane of the galaxy, $w(r, \theta)$ is the radial velocity in the plane of the galaxy and $z(r, \theta)$ is the

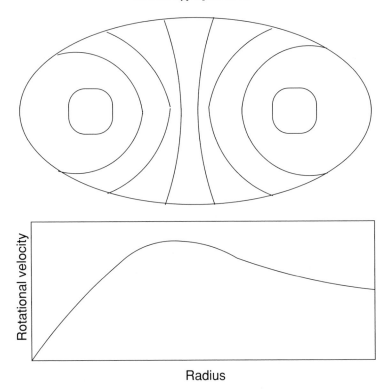

Fig. 14.3 Model of a Keplerian rotation curve expected from a thin circular disk, tilted at an angle i with respect to the line of sight. The top view shows isovelocity contours for the tilted disk. The bottom plot shows the radial velocity curve obtained by azimuthal integration of the velocity.

velocity out of the plane. For purely circular rotation, w and $z = 0$ so that

$$v(r)\cos\theta = \text{constant}$$

represents the lines of constant observed radial velocity along the line of sight. Azimuthal averaging produces the rotation curve shown in Fig. 14.3. Figure 14.4 shows the HI emission of the galaxy M33. The radial distribution of the HI is shown in Fig. 14.5. The velocity field is displayed in Fig. 14.6 and the corresponding rotation curve is shown in Fig. 14.7.

14.5 Probing galactic mass distributions

In this Section we will discuss the use of spectral lines to determine the mass distributions within galaxies as well as the distribution of matter on cosmological scales. The first part deals with masses of individual galaxies and addresses the

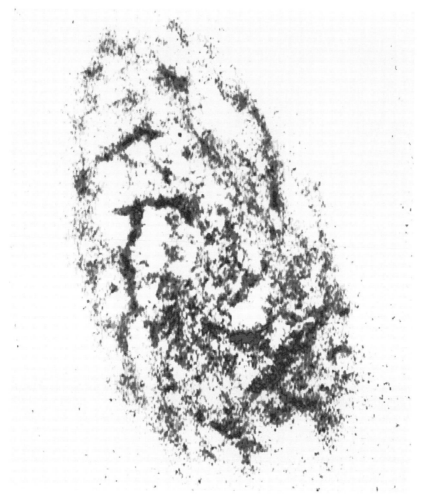

Fig. 14.4 The HI emission of M33. The intensity of the 21 cm line emission is shown in a grey-scale representation such that intensities increase with shade darkness. From: Wright, M. C. H., Warner, P. J., Baldwin, J. E., 1972, ApJ, 155, pp. 337–356.

issue of missing mass and dark matter. The second part deals with the distribution of HI on scales that can help us understand the evolution of the Universe as a whole.

14.5.1 *HI rotation curves*

For the purpose of interpreting rotation curves of galaxies, let us make the simplifying assumption that spiral galaxies have thin circular disks and central spherical bulges such that R_B, R_d and h represent the bulge radius, disk radius and disk thickness respectively (Fig. 14.8).

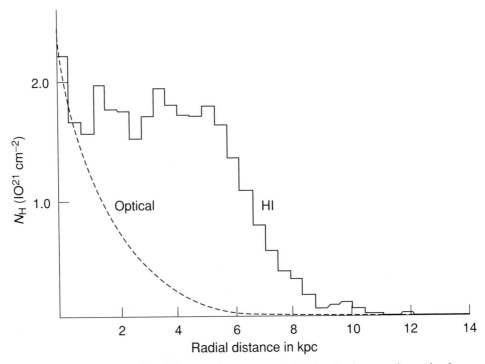

Fig. 14.5 The radial distribution of HI in M33. The HI distribution was determined through azimuthal integration of the emission map shown in Fig. 14.4 and the use of equation (445). The distribution of optical light is shown for comparison. From: Wright, M. C. H., Warner, P. J., Baldwin, J. E., 1972, ApJ, 155, pp. 337–356.

It is assumed that mass elements in the disk of a spiral galaxy have circular orbits about the center of the galaxy.

The bulge

If the bulge has a constant density ρ_B, independent of radius, then

$$M(r) = \frac{4}{3}\pi r^3 \rho_B \qquad (r < R_B)$$

represents the mass distribution as a function of the radius, r, out to the edge of the bulge, $r = R_B$. From (23) and the following, we have

$$v^2(r) = \frac{GM(r)}{r} = \frac{4}{3}\pi \rho_B G r^2 \qquad (r < R_B).$$

Thus, we see that

$$v(r) \propto r \qquad \text{and} \qquad \dot{\theta}(r) = \text{constant}$$

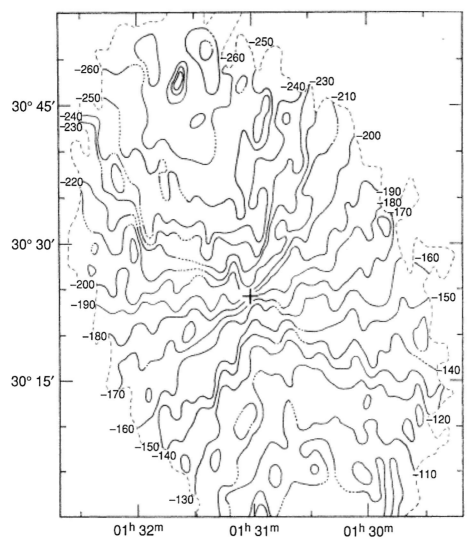

Fig. 14.6 The radial velocity field of M33. The observed velocity field is determined from measurements of the Doppler shift of the HI line and the use of equation (448). From: Warner, P. J., Wright, M. C. H., Baldwin, J. E., 1973, MNRAS, 163, pp. 163–182.

the linear velocity increases with r and the angular velocity is constant indicating rigid body rotation. This kind of rotation is observed in all spiral galaxy bulges and is consistent with the rotation of the inner disk as shown in Fig. 14.3.

The disk

Suppose that a point is reached where most of the disk is encompassed within an orbit of an outer mass element. Then we expect that mass element to respond as if

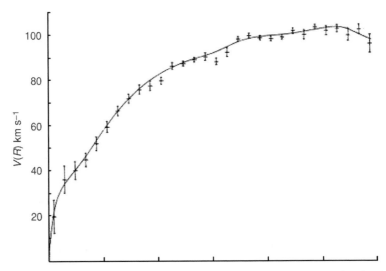

Fig. 14.7 The rotation curve obtained from the velocity field shown in Fig. 14.6. From: Warner, P. J., Wright, M. C. H., Baldwin, J. E., 1973, MNRAS, 163, pp. 163–182.

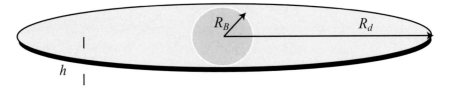

Fig. 14.8 The bulge and disk components of a spiral galaxy.

it were orbiting a point of mass. In the limit of large r equation (63) yields

$$v(r) \propto \frac{1}{\sqrt{r}} \quad \text{and} \quad \dot{\theta} \propto r^{-3/2}$$

consistent with $M(r) = $ constant and Keplerian rotation. These are indicative of *differential rotation*, a characteristic shared by many astronomical bodies that are not solid or rigid.

Thus, it is expected that in the outer-most regions of a spiral galaxy there should be a turnover where $v(r)$ begins to decline according to Keplerian orbital motion. Figure 14.9 shows a sample of *observed* rotation curves. It is evident that the expected turnover is rarely observed in spiral galaxies! In fact, most rotation curves appear to be flat, that is $v(r) = $ constant. Why is that? One explanation is that the $\rho(r) \propto 1/r$ which would yield $v(r) = $ constant at large radii (see previous equations). If that were the case we would expect that the surface brightness of a galaxy would also fall off as $1/r$. Figure 14.10 shows the typical surface brightness profile

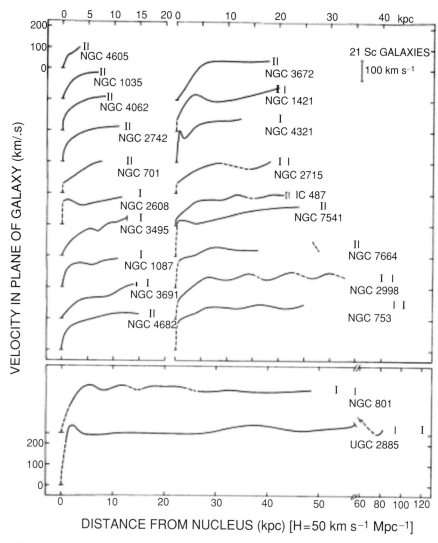

Fig. 14.9 Observed rotation curves. From: Rubin, V. C., Thonnard, N., Ford, W. K. Jr., 1980, ApJ, 238, pp. 471–487.

of a galaxy (in this case the spiral galaxy NGC 5055). The surface brightness falls off exponentially, too steeply to produce the required $1/r$ density law. Exponential surface brightness profiles characterize almost all spiral galaxies. The explanation must lie somewhere else, but where? Well, if we assume that Kepler's laws are valid on kiloparsec scales then we must conclude that the true mass distribution is different from that which is actually visible. The unseen mass may not radiate but it still provides a gravitational potential to which any mass element must respond.

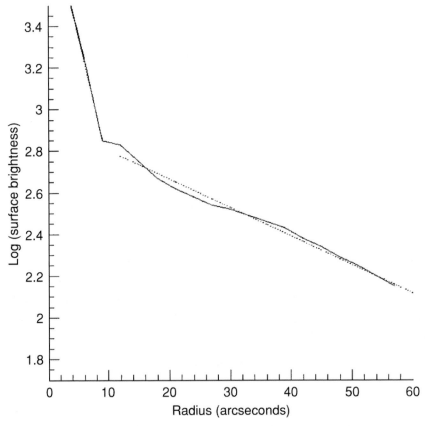

Fig. 14.10 The surface brightness profile of the galaxy NGC 5055. Note the bulge and disk components.

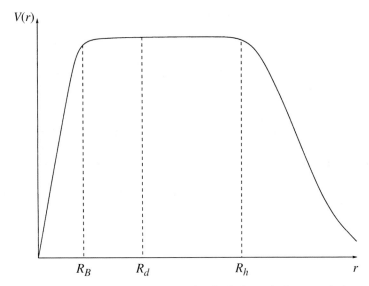

Fig. 14.11 A hypothetical rotation curve that includes a dark matter halo.

The halo

Let us therefore suppose that there is a massive, spherical and invisible halo of material surrounding a spiral galaxy such that

$$M(r) = \frac{4}{3}\pi r^3 \rho_h \qquad (r < R_h)$$

and let us suppose that the halo dominates all other components in terms of total mass. It turns out that such massive halos are gravitationally stable if $\rho_h \propto 1/r^2$ in which case

$$M(r) \propto r$$

$$v^2(r) \propto \frac{M(r)}{r} = \text{constant}$$

$$\Rightarrow v(r) = \text{constant}.$$

We see that massive halos of dark matter can *flatten* rotation curves of spiral galaxies. This fact is the main contention that individual galaxies are surrounded by dark matter. In a following chapter we discuss evidence for dark matter in clusters of galaxies. If these dark matter halos exist then the true rotation curves of galaxies might look something like the sketch in Fig. 14.11.

14.6 References

Jackson, J. D. (1998) *Classical Electrodynamics*, John Wiley and Sons, New York, NY, USA.

Rubin, V. C., Thonnard, N., Ford, W. K. Jr. (1980) Rotational Properties of 21 Sc Galaxies with a Large Range of Luminosities and Radii, *Astrophysical Journal*, **238**, pp. 471–487.

14.7 Further reading

Cohen-Tanoudji, C. (1992) *Principles of Quantum Mechanics*, John Wiley and Sons, New York, NY, USA, pp. 16–47.

Dirac, P. A. M. (1982) *Principles of Quantum Mechanics*, 4th edn, Oxford University Press, Oxford, UK.

Duric, N. *et al.* (1996) Radial Light Distributions in Five Spiral Galaxies: The Relationship between Young Stars and the Relativistic Gas, *Astrophysical Journal*, **470**, pp. 814–20.

Griffiths, D. (1995) *Introduction to Quantum Mechanics*, Prentice Hall, Englewood Cliffs, NJ, USA.

Landau, L. D. and Lifschitz, E. M. (1977) *Quantum Mechanics*, 3rd edn, Pergamon Press, Oxford, UK.

Landau, R. H. (1996) *Quantum Mechanics II: A Second Course in Quantum Theory*, 2nd edn, John Wiley and Sons, New York, NY, USA.

Osterbrock, D. E. (1974) *Astrophysics of Gaseous Nebulae*, W. H. Freeman, San Francisco, CA, USA.

Shu, F. H. (1992) *The Physics of Astrophysics: Radiation*, University Science Books, Mill Valley, CA, USA.

Verschuur, G. A. and Kellerman, K. I. (1988) *Galactic and Extragalactic Radio Astronomy*, Springer-Verlag, Berlin, Germany.

Chapter 15

Transitions involving multi-electron atoms

Thus far, I have discussed only hydrogenic atoms, those with a single electron. In order to expand our understanding to atoms with multiple electrons we will do two things. First, we will discuss the necessary issue of the symmetry of multi-particle wave functions. Second, we will consider the structure of the helium atom in considerable detail. Understanding the helium atom lays the foundation for understanding the general case of the multi-electron atom.

15.1 Symmetry of multi-particle wave functions

Recall that there are four ways to construct a two-particle spin wave function from the individual spin wave functions.

$$
\chi = \begin{cases}
\chi_+(1)\chi_+(2) \\
\chi_-(1)\chi_-(2) \\
\frac{1}{\sqrt{2}}(\chi_+(1)\chi_-(2) + \chi_-(1)\chi_+(2)) \\
\frac{1}{\sqrt{2}}(\chi_+(1)\chi_-(2) - \chi_-(1)\chi_+(2)).
\end{cases}
\tag{449}
$$

The first three functions are said to be *symmetric* because the interchange of particles 1 and 2 does not lead to a change in sign of χ. The fourth function is *antisymmetric* because the interchange leads to a change in sign of χ.

The total wave function for an electron, $\psi_n = \phi_n\chi_\pm$, is the product of the spatial and spin wave functions. Thus, the total wave function of two electrons can be defined as the product of the two-electron spatial function and the two-electron spin function.

$$
\psi = \phi\chi = \frac{1}{\sqrt{2}}\left[\phi_n(1)\phi_{n'}(2) \pm \phi_{n'}(1)\phi_n(2)\right]\chi
$$

where χ is given by (449). We see that there are potentially eight possible wave functions (possibly more if n and n' are degenerate).

In the case of electrons, being fermions they must satisfy the Pauli exclusion principle which states that no two electrons can occupy the same state. Thus, if the two electrons are in the same spatial state they must be in different spin states. This is another way of saying that *the total wave function must be antisymmetric*. Thus, if the spatial two-electron wave function is symmetric, the two-electron spin function must be antisymmetric and if the spatial function is antisymmetric the spin function must be symmetric. According to these constraints the above functions become

$$
\psi = \frac{1}{\sqrt{2}}
\begin{cases}
[\phi_n(1)\phi_{n'}(2) - \phi_{n'}(1)\phi_n(2)]
\begin{cases}
\chi_+(1)\chi_+(2) \\
\chi_-(1)\chi_-(2) \\
\frac{1}{\sqrt{2}}[\chi_+(1)\chi_-(2) + \chi_-(1)\chi_+(2)]
\end{cases} \\[2em]
[\phi_n(1)\phi_{n'}(2) + \phi_{n'}(1)\phi_n(2)]\left[\frac{1}{\sqrt{2}}(\chi_+(1)\chi_-(2) - \chi_-(1)\chi_+(2))\right].
\end{cases}
$$

There are four possible two-electron wave functions that satisfy the Pauli exclusion principle. We are now ready to discuss the helium atom.

15.2 The helium atom

The helium atom consists of two electrons and a nucleus made up of two protons and two neutrons. The two electrons interact with the nucleus via the Coulomb potential. In addition, the electrons feel a Coulomb repulsion between themselves, and one electron always screens some of the nuclear charge that the other electron "sees". Finally, the constraints imposed by the Pauli exclusion principle, that is the required antisymmetry of the two-electron wave function, introduce a subtle force-like effect called the *exchange force* or sometimes called the *correlation force*. All these effects (plus other even more subtle ones from quantum field theory) act together to determine the structure of the helium atom. We now describe the ground and first excited state of helium in light of these various effects.

15.2.1 The ground state

We begin with the strongest effect, that of the electron Coulomb interaction with the nucleus. In that case we can write the Hamiltonian as

$$
H = \left(-\frac{\hbar^2}{2m}\nabla_1^2 - \frac{2e^2}{r_1}\right) + \left(-\frac{\hbar^2}{2m}\nabla_2^2 - \frac{2e^2}{r_2}\right) = 2H_0. \qquad (Z = 2).
$$

The subscripts 1 and 2 refer to the electrons 1 and 2. For a hydrogenic atom

$$
H_0 = -Z^2 E_0 = -54.4 \text{ eV}.
$$

Thus, for helium

$$H_{He} = 2H_0 = -108.8 \text{ eV} \qquad (450)$$

some way from the measured value of -78.62 eV.

Clearly, we need to add the electron–electron Coulomb repulsion because it will raise the total energy of the system. Treating the electron–electron repulsion as a perturbation we can modify the Hamiltonian so that

$$H = 2H_0 + \frac{e^2}{r_{12}} \qquad r_{12} = |\vec{r}_1 - \vec{r}_2|.$$

We can now define the energy of the ground state by taking the expectation value of H, so that

$$\langle 0|H|0\rangle = -2Z^2 E_0 + \left\langle 0 \left| \frac{e^2}{r_{12}} \right| 0 \right\rangle$$

where

$$\left\langle 0 \left| \frac{e^2}{r_{12}} \right| 0 \right\rangle = \frac{e^2}{\pi^2} \left(\frac{Z}{a_0} \right)^6 \left\langle e^{-(Z/a_0)(r_1+r_2)} \left| \frac{1}{r_{12}} \right| e^{-(Z/a_0)(r_1+r_2)} \right\rangle \langle s, m_s | s, m_s \rangle$$

$$= \frac{e^2}{\pi^2} \left(\frac{Z}{a_0} \right)^6 \int_{\Omega_1} d\Omega_1 \int_0^\infty dr_1 r_1^2 e^{-2Zr_1/a_0} \int_{\Omega_2} d\Omega_2 \int_0^\infty dr_2 \frac{r_2^2}{r_{12}} e^{-2Zr_2/a_0}$$

$$= \frac{5e^2 Z}{8a_0} = \frac{5}{4} Z E_0 \qquad (451)$$

Adding (450) to (451), we get the modified energy of the ground state

$$E_g = -2Z^2 E_0 + \frac{5}{4} Z E_0 = -\frac{11}{2} E_0 = -74.8 \text{ eV}.$$

This is closer! Now let us consider a further correction associated with the partial screening of the nucleus by the electrons. This can be done with a variational approach in which we try to minimize the energy with respect to the effective value of Z which is allowed to vary as a free parameter.

The expectation value of the ground state energy can be written as

$$\left\langle 0 \left| 2H_0 + \frac{e^2}{r_{12}} \right| 0 \right\rangle = 2\langle 0|H_0|0\rangle + \left\langle 0 \left| \frac{e^2}{r_{12}} \right| 0 \right\rangle. \qquad (452)$$

The latter we have already calculated. Evaluation of the former leads to

$$2\langle 0|H_0|0\rangle = 2(Z^2 - 4Z)E_0.$$

Substituting into (452) yields

$$\langle E_g\rangle = 2(Z^2 - 4Z)E_0 + \frac{5}{4}ZE_0. \tag{453}$$

All we have to do now is minimize the above and solve for the Z at the minimum. Thus, setting $(d/dZ)\langle E_g\rangle = 0$ we get

$$Z = \frac{27}{16}.$$

Substituting into (453) leads to the newest estimate of the ground state energy

$$\langle E_g\rangle = -5.7E_0 = -77.5 \text{ eV}.$$

15.2.2 *Lowest excited states*

The lowest possible excited state occurs when one electron remains in the ground (1s) state while the other is excited into the 2s or 2p state. As discussed earlier, for a pair of electrons, there are four possible spin wave functions and in this case four possible spatial wave functions (the $\psi_{100}\psi_{200}$, $\psi_{100}\psi_{210}$, $\psi_{100}\psi_{21,\pm1}$ states). Thus, there are sixteen possible states. The Coulomb interactions lift some of the degeneracy of these states. I now demonstrate this explicitly. Let us begin by writing the two-electron wave functions for the 1s, 2s excited state as

$$\psi = \frac{1}{\sqrt{2}}[\psi_{1s}(1)\psi_{2s}(2) \pm \psi_{2s}(1)\psi_{1s}(2)] \, s, m_s\rangle.$$

The first order correction to E_2 is then given by

$$E_2^{(1)} = \left\langle \psi \left| \frac{e^2}{r_{12}} \right| \psi \right\rangle$$

$$= \frac{e^2}{2}\left\langle \psi_{1s}(1)\psi_{2s}(2) \pm \psi_{2s}(1)\psi_{1s}(2) \left| \frac{1}{r_{12}} \right| \psi_{1s}(1)\psi_{2s}(2) \right.$$

$$\left. \pm \psi_{2s}(1)\psi_{1s}(2) \right\rangle\langle s, m_s|s, m_s\rangle$$

$$= K \pm J$$

where

$$K = e^2 \left\langle \psi_{1s}(1)\psi_{2s}(2) \left| \frac{1}{r_{12}} \right| \psi_{1s}(1)\psi_{2s}(2) \right\rangle$$

and

$$J = e^2 \left\langle \psi_{1s}(1)\psi_{2s}(2) \left| \frac{1}{r_{12}} \right| \psi_{1s}(2)\psi_{2s}(1) \right\rangle.$$

The K correction is known as the ordinary Coulomb integral while J is the exchange Coulomb integral and represents the so-called *exchange force*.

The first order correction to E_2 is therefore given by

$$E_2 = E_2^{(0)} + K \pm J.$$

The K correction acts as a d.c. offset, raising the energy of the level by a fixed amount. The J correction (from the exchange force) actually splits the level into two. The antisymmetric spin state is raised while the triplet of symmetric spin states is lowered with respect to K. A similar effect occurs in the case of the 1s, 2p state, although in this case the value of K is higher. The net effect is to split the Bohr $n = 2$ level into four levels. The values of K and J associated with the various splittings are summarized in Fig. 15.1.

$$K_{1s,2p} = e^2 \left\langle \psi_{1s}(1)\psi_{2p}(2) \left| \frac{1}{r_{12}} \right| \psi_{1s}(1)\psi_{2p}(2) \right\rangle \approx 10.0 \text{ eV}$$

$$K_{1s,2s} = e^2 \left\langle \psi_{1s}(1)\psi_{2s}(2) \left| \frac{1}{r_{12}} \right| \psi_{1s}(1)\psi_{2s}(2) \right\rangle \approx 9.1 \text{ eV}$$

$$J_{1s,2s} = e^2 \left\langle \psi_{1s}(1)\psi_{2s}(2) \left| \frac{1}{r_{12}} \right| \psi_{1s}(2)\psi_{2s}(1) \right\rangle \approx 0.4 \text{ eV}$$

$$J_{1s,2p} = e^2 \left\langle \psi_{1s}(1)\psi_{2p}(2) \left| \frac{1}{r_{12}} \right| \psi_{1s}(2)\psi_{2p}(1) \right\rangle \approx 0.1 \text{ eV}.$$

15.2.3 Summary

The basic interactions that lead to the structure of the helium atom are Coulomb interactions combined with the Pauli exclusion principle, the latter leading to the exchange force. There is additional structure arising from the fact that the splittings are affected by the orbital angular momentum of the electrons and the fine and hyperfine structure already discussed in the case of single-electron atoms. It must be remembered that in many cases, when dealing with multi-electron atoms, the Coulomb interactions and the exchange force lead to greater splittings than spin–orbit coupling. These effects cannot be ignored, in general, for multi-electron atoms. There is a methodology that describes the structure of multi-electron atoms, which incorporates all these effects. It is called the *LS coupling* or *Russell–Saunders coupling* scheme.

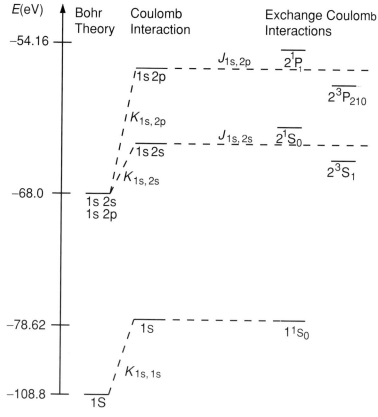

Fig. 15.1 Energy levels of the helium atom. The diagram shows the splittings arising from Coulomb and exchange interactions.

15.3 Many-electron atoms

The LS coupling scheme is based on the following sequence of "corrections" to a level of a given principal quantum number. These are known as *Hund's Rules.*

1. The higher the S (total electron spin), the lower the energy. These splittings arise from the exchange force.
2. The higher the L (total electron orbital momentum), the lower the energy of the exchange splitting. This results from the fact that at high orbital momenta the electrons "keep apart" better thereby enhancing the effect of the exchange force.
3. The higher the J the higher the energy associated with spin–orbit splitting. This is true for electron shells that are less than half filled. For shells that are more than half filled the opposite is true.

Figure 15.2 summarizes these results schematically.

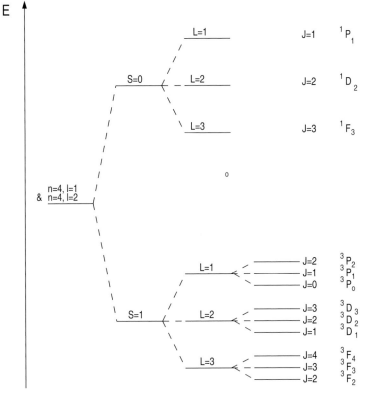

Fig. 15.2 The LS coupling scheme. The splittings for the $n = 4$ level are shown.

15.3.1 The Hartree–Fock procedure

To actually calculate the energies of all the split levels it is necessary to invoke either numerical procedures or analytical approximations. The Hartree–Fock procedure utilizes both.

For an N-electron atom we can express the Hamiltonian as

$$H = H_{cf} + H_{ee} + H_{SO}$$

where H_{cf} represents the basic central field approximation (assume each electron acts as if it is in a hydrogenic atom) so that

$$H_{cf} = \sum_{a=1}^{N} H_a$$

where

$$H_a \equiv -\frac{\hbar^2}{2m}\nabla_a^2 + V_a(r_a)$$

and

$$V_a(r_a) = -\frac{Ze^2}{r_a} + W_a(r_a).$$

The function $W_a(r_a)$ is meant to be a spherical approximation for a smeared-out electron cloud distribution.

The second term, H_{ee}, represents the electron–electron repulsion in excess of any spherically symmetric effects (which cancel out),

$$H_{ee} = \frac{e^2}{2} \sum_{a\neq b}^{N} \frac{1}{r_{ab}} - \sum_{a=1}^{N} W_a(r_a).$$

Finally, H_{SO} represents the spin–orbit interaction

$$H_{SO} = \sum_{a=1}^{N} \frac{1}{2m^2c^2} \left(\frac{1}{r_a}\frac{dV_a}{dr_a}\right) \vec{S}_a \cdot \vec{L}_a.$$

One proceeds by constructing an antisymmetric wave function using combinations of unperturbed single-particle wave functions. The functions are taken to be trial functions. The expectation of H is then minimized with respect to the trial functions, which yields an estimate of the wave function.

15.4 Forbidden lines in astrophysics

Some multi-electron atoms have low lying energy levels that are easier to excite collisionally than in the case of the hydrogen atom. Examples include OII, NII, SII and OIII (Fig. 15.3). The first few energy levels of these atoms have energies comparable to the kT of most astrophysical nebulas and even active galactic nuclei (such as Seyferts) and quasars. Thus, the spectral lines associated with these low-lying energy levels are ideal diagnostics of the physical conditions in these regions. In this portion of the chapter we will describe how density and temperature are measured from ratios of spectral line intensities.

15.4.1 Collisional equilibria

We begin by considering the process of collisional excitation and de-excitation of the low lying energy levels. In analogy to radiative cross-sections we discussed earlier we define the collision cross-section as

$$\sigma_{12}(v) = \frac{\pi\hbar^2}{m^2v_1^2}\frac{\Omega_{12}}{g_1} \qquad \frac{1}{2}mv^2 > \chi_{12} \tag{454}$$

where Ω_{12} is the quantum mechanical collision strength and is approximately constant near the threshold energy. The statistical weight, $g_1 = 2J(1) + 1$, is the statistical weight of the level being depopulated. The energy gap between the two levels is given by χ_{12}, the threshold energy.

The impacting particles tend to be electrons because they have the highest cross-section for interacting with ions. Since the nebulas are usually thermal in character we can describe the electron velocity distribution as Maxwellian. Consider a 2-level system with $n = 1, n = 2$. There is a balance between collisions which is such that for every excitation there is a de-excitation

$$N_e N_1 v_1 \sigma_{12}(v_1) f(v_1) \, dv_1 = N_e N_2 v_2 \sigma_{21} f(v_2) \, dv_2 \tag{455}$$

where N is the population of each level, $\Delta v = v_1 + dv_1$ is the velocity range for an electron producing an excitation, $\Delta v = v_2 + dv_2$ for an electron producing a de-excitation. Combining with the Boltzmann relation

$$\frac{N_1}{N_2} = \frac{g_2}{g_1} e^{-\chi_{12}/kT} \tag{456}$$

we can derive the relation

$$g_1 v_1^2 \sigma_{12}(v_1) = g_2 v_2^2 \sigma_{21}(v_2) \tag{457}$$

where $(1/2)mv_1^2 = (1/2)mv_2^2 + \chi_{12}$ so that $v_1 \, dv_1 = v_2 \, dv_2$.
    Combining (454) and (457) yields

$$\sigma_{21}(v_2) = \frac{\pi \hbar^2}{m v_2^2} \frac{\Omega_{12}}{g_2} \tag{458}$$

and we see that $\Omega_{12} = \Omega_{21}$. We are now in the position of being able to calculate the total rate of collisional de-excitations per unit volume

$$N_e N_2 q_{21} = N_e N_2 \int_0^\infty \sigma_{21}(v) v f(v) \, dv$$

$$= N_e N_2 \left(\frac{2\pi}{kT}\right)^{1/2} \frac{\hbar^2}{m^{3/2}} \frac{\Omega_{12}}{g_2} = \frac{N_e N_2 8.6 \times 10^{-6}}{T^{1/2}} \frac{\Omega_{12}}{g_2} \tag{459}$$

for a constant Ω_{12} and units of cm^3 s^{-1}. The rate of excitations is similarly given by $N_e N_1 q_{12}$ where

$$q_{12} = \frac{g_2}{g_1} q_{21} e^{-\chi_{12}/kT}. \tag{460}$$

Tables 15.1 and 15.2 list the collisional strengths for a variety of ions and transitions. We now examine, in light of the previous discussion, how spectral lines are formed in astrophysical nebulas.

Table 15.1. *Collisional strengths for doubly ionized atoms*

Ion	$\Omega(^3P, ^1D)$	$\Omega(^3P, ^1S)$	$\Omega(^1D, ^1S)$	$\Omega(^3P_0, ^3P_1)$	$\Omega(^3P_0, ^3P_2)$	$\Omega(^3P_1, ^3P_2)$
NII	2.99	0.36	0.39	0.41	0.28	1.38
OIII	2.50	0.30	0.58	0.39	0.21	0.95
SIII	3.87	0.75	1.34	0.94	0.51	2.32

Table 15.2. *Collisional strengths for singly ionized atoms*

Ion	$\Omega(^4S, ^2D)$	$\Omega(^4S, ^2P)$	$\Omega(^2D_{3/2}, ^2D_{5/2})$	$\Omega(^2D_{3/2}, ^2P_{1/2})$
OII	1.47	0.45	1.16	0.29
SII	5.66	2.72	5.55	1.96

15.4.2 Line emission and cooling of nebulas

Consider a simple two-level ion in the limit of low density. Every collisional excitation is followed by a radiative de-excitation. The line emissivity is then given by

$$j_{21} = N_2 A_{21} h\nu_{21} = N_e N_1 q_{12} h\nu_{21} \tag{461}$$

since

$$N_2 A_{21} = N_e N_1 q_{12}.$$

When the density is significant enough to make collisional de-excitation important the excitation balance becomes

$$N_e N_1 q_{12} = N_e N_2 q_{21} + N_2 A_{21} \tag{462}$$

so that

$$\frac{N_2}{N_1} = \frac{N_e q_{12}}{A_{21}} \left[\frac{1}{1 + N_e q_{21}/A_{21}} \right] \tag{463}$$

and the line emissivity now becomes

$$j_{21} = N_2 A_{21} h\nu_{21} = N_e N_1 q_{12} h\nu_{21} \left[\frac{1}{1 + N_e q_{21}/A_{21}} \right]. \tag{464}$$

If we let $N_e \to 0$ we recover (461). If we let $N_e \to \infty$ we get

$$j_{21} \to N_1 \frac{q_{12}}{q_{21}} h\nu_{21} A_{21}.$$

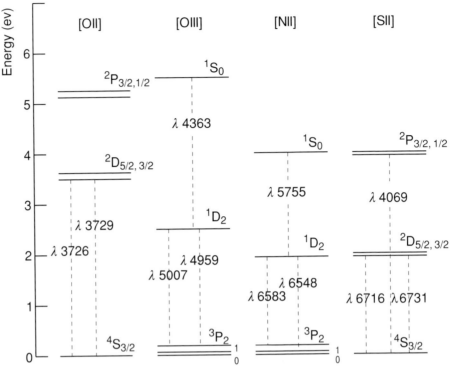

Fig. 15.3 Energy levels of OII, OIII, SII and NII. Note the low lying states that are only 2–3 eV above the ground state. These states can be populated collisionally.

Combining with (460) yields

$$j_{21} \rightarrow N_1 \frac{g_2}{g_1} e^{-\chi_{12}/kT} h\nu_{21} A_{21} \qquad (465)$$

which is what we would expect under conditions of thermodynamic equilibrium (in the limit of high density). Examination of (461) and (465) shows that at low densities the strength of a line grows as N^2 while at high densities it grows as N. Also, at low densities there is no dependence on A_{21}. This means that forbidden lines are treated on an equal footing with the permitted lines. However, at high densities the forbidden lines are "quenched" by the ratio of the permitted to forbidden transition rates, a ratio that is at least six orders of magnitude smaller than 1. The levels that would normally be populated for the forbidden lines are depopulated by the combination of collisions and permitted transitions. Figure 15.3 shows the lowest levels of OII, OIII, SII and NII.

Only the bottom five levels are normally active since their energies are near $kT \approx 1\,\text{eV}$, typical of HII regions and other nebulas. Note that all possible transitions in the optical violate the dipole selection rules for multi-electron atoms. Thus all

Table 15.3. *Transition rates for O[II] and S[II]*

	O[II]		S[II]	
Transition	$A\,(\mathrm{s}^{-1})$	$\lambda\,(\mathring{\mathrm{A}})$	$A\,(\mathrm{s}^{-1})$	$\lambda\,(\mathring{\mathrm{A}})$
$^2P_{1/2}-^3P_{3/2}$	6.0×10^{-11}	–	1.0×10^{-6}	–
$^2D_{5/2}-^2P_{3/2}$	1.2×10^{-1}	7319.9	2.1×10^{-1}	10320.5
$^2D_{3/2}-^2P_{3/2}$	6.1×10^{-2}	7330.2	1.7×10^{-1}	10286.7
$^2D_{5/2}-^2P_{1/2}$	6.1×10^{-2}	7319.9	8.7×10^{-2}	10370.5
$^2D_{3/2}-^2P_{1/2}$	1.0×10^{-1}	7330.2	2.0×10^{-1}	10336.4
$^4S_{3/2}-^2P_{3/2}$	6.0×10^{-2}	2470.3	3.4×10^{-1}	4068.6
$^4S_{3/2}-^2P_{1/2}$	2.4×10^{-2}	2470.3	1.3×10^{-1}	4076.4
$^2D_{5/2}-^2D_{3/2}$	1.3×10^{-7}	–	3.3×10^{-7}	–
$^4S_{3/2}-^2D_{5/2}$	4.2×10^{-5}	3728.8	4.7×10^{-4}	6716.4
$^4S_{3/2}-^2D_{3/2}$	1.8×10^{-4}	3726.1	1.8×10^{-3}	6730.8

Table 15.4. *Transition rates for N[II], O[III] and S[III]*

	N[II]		O[III]		S[III]	
Transition	$A\,(\mathrm{s}^{-1})$	$\lambda\,(\mathring{\mathrm{A}})$	$A\,(\mathrm{s}^{-1})$	$\lambda\,(\mathring{\mathrm{A}})$	$A\,(\mathrm{s}^{-1})$	$\lambda\,(\mathring{\mathrm{A}})$
$^1D_2-^1S_0$	1.1	5754.6	1.6	4363.2	2.5	6312.1
$^3P_2-^1S_0$	1.6×10^{-4}	3070.8	7.1×10^{-4}	2331.4	1.6×10^{-2}	3797.4
$^3P_1-^1S_0$	3.4×10^{-2}	3062.8	2.3×10^{-1}	2321.0	8.5×10^{-1}	3721.7
$^3P_2-^1D_2$	3.0×10^{-3}	6583.4	2.1×10^{-2}	5006.9	6.4×10^{-2}	9531.8
$^3P_1-^1D_2$	1.0×10^{-3}	6548.1	7.1×10^{-3}	4958.9	2.5×10^{-2}	9069.0
$^3P_0-^1D_2$	4.2×10^{-7}	6527.1	1.9×10^{-6}	4931.0	9.1×10^{-6}	8830.9
$^3P_1-^3P_2$	7.5×10^{-6}	–	9.8×10^{-5}	–	2.4×10^{-3}	–
$^3P_0-^3P_2$	1.3×10^{-12}	–	3.5×10^{-11}	–	4.7×10^{-8}	–
$^3P_0-^3P_1$	2.1×10^{-6}	–	2.6×10^{-5}	–	4.7×10^{-4}	–

transitions we will be considering are forbidden. Tables 15.3 and 15.4 list the transition rates for the more common spectral lines of multi-electron ions.

Let us now discuss the OIII ion and ions like it. Since these are not simple two-level systems as previously discussed we need to examine the general equation of statistical equilibrium.

15.4.3 *Statistical equilibrium for N levels*

For each level i of a multi-level atom it is necessary to consider all possible paths for populating and depopulating level i. Thus

$$\sum_{j\neq i} N_j N_e q_{ji} + \sum_{j>i} N_j A_{ji} = \sum_{j\neq i} N_i N_e q_{ij} + \sum_{j<i} N_i A_{ij}. \qquad (466)$$

Table 15.5. *Critical densities*

Ion	Level	N_e (cm^3)
NII	$^1D_2$	7.8×10^4
NII	$^3P_2$	2.6×10^2
NII	$^3P_1$	41
OII	$^2D_{3/2}$	3.3×10^3
OII	$^2D_{5/2}$	6.3×10^2
OIII	$^1D_2$	6.5×10^5
OIII	$^3P_2$	4.9×10^3
OIII	$^3P_1$	6.7×10^2

The above is subject to number conservation, $\sum_j N_j = N$. Equation (466) can be solved to yield equilibrium populations for any level i. Once the populations are determined the line emissivity follows from

$$j_{ik} = N_i A_{ik} h \nu_{ik}.$$

The above is greatly simplified in the low density limit because j_{ik} can be expressed as terms of the form

$$N_i N_e q_{ki} h \nu_{ik}$$

making the previous equation a lot easier to solve. The low density approximation is valid so long as

$$N_e < N_c = A_{ik}/q_{ik}.$$

For typical interstellar conditions, $N_c \approx 10^5$ cm^{-3}. Table 15.5 lists critical densities for a variety of ions and transitions.

For most nebulas the densities are below the critical density and the low density approximation can often be used. We will now use that approximation to discuss how OIII lines can be used to estimate temperatures of astrophysical gases.

15.4.4 OIII lines as probes of temperature

The most common way of estimating temperatures of nebulas is to use the ratio of spectral line intensities such as

$$\frac{j(\lambda 5007) + j(\lambda 4959)}{j(\lambda 4363)} = \frac{j(^1D_2 \rightarrow {}^3P_2) + j(^1D_2 \rightarrow {}^3P_1)}{j(^1S_0 \rightarrow {}^1D_2)}$$

where the corresponding transitions are indicated. Note that all three transitions violate the dipole selections rules and are therefore forbidden. Can you find *any*

permitted transitions in the OIII energy level diagram? Let us now set up an excitation balance equation for each transition in the low density limit.

For the doublet in the numerator

$$N_e N_1 q_{12} = N_e [N_{1a} q_{21a} + N_{1b} q_{21b} + N_{1c} q_{21c}]$$
$$= N_2 A_{21} = N_2 [A_{21a} + A_{21b} + A_{21c}]$$

where the radiative population of the $n = 2$ level is negligible. We can further
simplify this by noting that the collisional excitations from the ground triplet are
more or less independent of level (because the ΔE is essentially the same for all
three) so that an average q_{12} can be used. The downward $^1D_2 \rightarrow {}^3P_x$ transitions
are represented by the subscripts a, b and c for $x = 1, 2, 3$ respectively. Since
$A_{21c} \ll A_{21b}, A_{21a}$ the equation reduces to

$$N_e N_1 q_{12} = N_2 [A_{21a} + A_{21b}].$$

Similar considerations with regard to the $^1S_0 \rightarrow {}^1D_2$ line yield

$$N_e N_1 q_{13} = N_3 [A_{31} + A_{32}].$$

The cross-section for photon absorption is proportional to A so that it is very
small for forbidden transitions and can be ignored. That, in fact, is the reason why
forbidden lines are important as coolants in nebulas allowing them to maintain an
equilibrium temperature. We are now in the position of evaluating the line ratio
presented above.

$$\frac{j(\lambda 5007) + j(\lambda 4959)}{j(\lambda 4363)} = \frac{N_2 [A_{21a} h \nu_{21a} + A_{21b} h \nu_{21b}]}{N_3 [A_{32} h \nu_{32}]}. \tag{467}$$

From the preceding discussion

$$N_2 = \frac{N_e N_1 q_{12}}{A_{21a} + A_{21b}}$$

$$N_3 = \frac{N_e N_1 q_{13}}{A_{31} + A_{32}}.$$

Substituting into (467) yields

$$\frac{j(\lambda 5007) + j(\lambda 4959)}{j(\lambda 4363)} = \frac{q_{12}}{q_{13}} \frac{A_{21a} h \nu_{21a} + A_{21b} h \nu_{21b}}{A_{21a} + A_{21b}} \frac{A_{31} + A_{32}}{A_{32} h \nu_{32}}.$$

Combining with (459) and (460) allows us to rewrite the above as

$$\frac{j(\lambda 5007) + j(\lambda 4959)}{j(\lambda 4363)} = \frac{\Omega_{21}}{\Omega_{31}} e^{\chi_{32}/kT} \frac{A_{21a} h \nu_{21a} + A_{21b} h \nu_{21b}}{A_{21a} + A_{21b}} \frac{A_{31} + A_{32}}{A_{32} h \nu_{32}}.$$

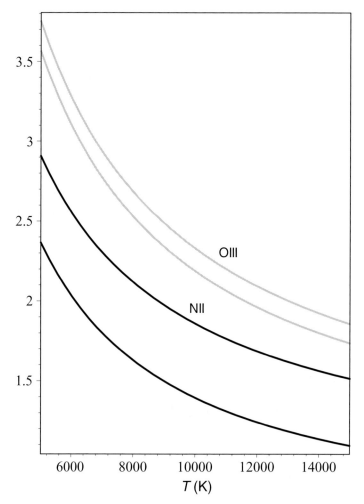

Fig. 15.4 Line ratios as functions of temperature for [OIII] and [NII]. For each element, the lower curve is associated with the low densities limit, while the upper curve represents the high density limit.

Using the A's and Ω's from Tables 15.1 to 15.4 we get

$$\frac{j(\lambda 5007) + j(\lambda 4959)}{j(\lambda 4363)} = 8.3 \, e^{3.3 \times 10^4 / T}.$$

We see that the line ratio is very sensitive to the temperature. Application of this procedure to measurements of nebulas yields temperatures generally in the range of 8×10^3–1.2×10^4 K. Figure 15.4 shows the dependence of the line ratio on temperature (along with an analogous ratio for NII).

15.4.5 Line ratios as density probes

Consider the energy level diagram of SII. The line ratio $j(\lambda 6716)/j(\lambda 6731)$ is potentially a useful estimator of density because it has almost no dependence on temperature since the two lines have almost identical excitation requirements. Allowing for any possible range of density we have to evaluate the ratio by considering excitation equilibria that include collisional de-excitation. Thus,

$$\frac{j(\lambda 6716)}{j(\lambda 6731)} = \frac{N_{2a} A_{2a1} h\nu_{2a1}}{N_{2b} A_{2b1} h\nu_{2b1}} \approx \frac{N_{2a} A_{2a1}}{N_{2b} A_{2b1}} \tag{468}$$

where the higher energy level in the doublet is labeled $2a$ and the lower level is $2b$, and where we have taken advantage of the fact that the ratio of photon energies is almost unity. From (463)

$$\frac{N_{2a}}{N_1} = \frac{N_e q_{12a}}{A_{2a1}} \left[\frac{1}{1 + N_e q_{2a1}/A_{2a1}} \right]$$

and

$$\frac{N_{2b}}{N_1} = \frac{N_e q_{12b}}{A_{2b1}} \left[\frac{1}{1 + N_e q_{2b1}/A_{2b1}} \right].$$

Combining these two equations with (468) we get

$$\frac{j(\lambda 6716)}{j(\lambda 6731)} = \frac{q_{12a}}{q_{12b}} \cdot \frac{1 + N_e q_{2b1}/A_{2b1}}{1 + N_e q_{2a1}/A_{2a1}} \tag{469}$$

which represents the relationship between N_e and the line ratio. We have ignored all other levels because they are generally too hard to excite for the range of temperatures typical of nebulas.

In the low density limit, $N_e \ll A_{2a1}/q_{2a1}$ so that

$$\frac{j(\lambda 6716)}{j(\lambda 6731)} = \frac{q_{12a}}{q_{12b}} = \frac{\Omega_{2a1}}{\Omega_{2b1}} e^{\chi_{2a2b}/kT} \approx \frac{\Omega_{2a1}}{\Omega_{2b1}} = \frac{g_{2a}}{g_{2b}}. \tag{470}$$

For the $^2D_{5/2}$ ($= 2a$) and $^2D_{3/2}$ ($= 2b$) levels, $g_{2a} = 6$ and $g_{2b} = 4$ so that the line ratio $\rightarrow 3/2$ in the low density limit.

In the high density limit, $N_e \gg A_{2a1}/q_{2a1}$ so that (469) reduces to

$$\frac{j(\lambda 6716)}{j(\lambda 6731)} = \frac{g_{2a}}{g_{2b}} \frac{A_{2a1}}{A_{2b1}} \approx 0.38$$

(see Tables 15.1 to 15.5). Figure 15.5 is a graphical representation of the dependence of the line ratios of OII and SII on density. We see that the line ratio is indeed sensitive, particularly in the vicinity of the critical density.

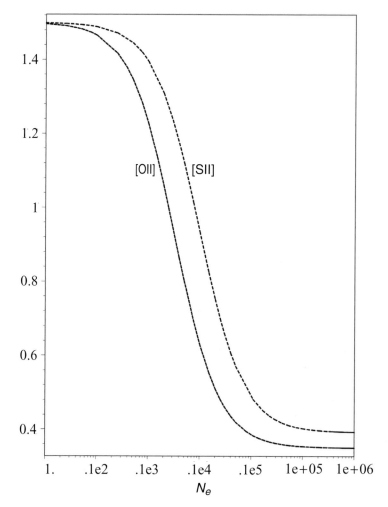

Fig. 15.5 Line ratios as functions of density for OII and SII. Note the great sensitivity to density in the range $N_e = 10-10^4$.

15.4.6 Observations of nebulas

Using techniques such as those previously described, the temperatures and densities of various astrophysical nebulas have been measured. Table 15.6 summarizes the properties of some well-known nebulas.

15.4.7 Observations of active galactic nuclei

Distant, active galaxies have very bright nuclei which can be detected to great distances. Analysis of the light from these active galactic nuclei (AGNs) can be performed in the same way as in the case of the nebulas. These measurements have

Table 15.6. *Physical parameters derived from observations*

Nebula	T_e (K)		T_k (K)	N_e (cm^{-3})	
	OIII	NII		OII	SII
Orion	8600	10 200	7000–12 000	100–3000	1000
Trifid	–	–	8000–10 000	100–200	300–1000
Lagoon	8300	8 100	7000–10 000	100–1000	1000–3000
Omega	9000	7 000	–	–	–

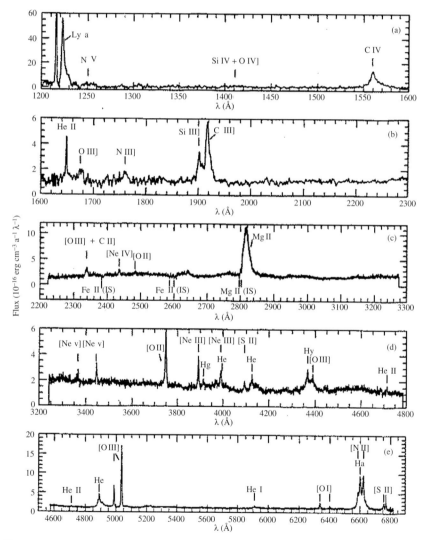

Fig. 15.6 Spectrum of a Seyfert galaxy. The various lines discussed in the preceding chapters can be seen here. Note the range from 1200 Å to 6900 Å.

led to a better understanding of the Seyfert and quasar phenomena. The spectrum shown in Fig. 15.6 illustrates the diverse spectral features found in AGNs.

15.5 Further reading

Avrett, E. H. (Ed) (1976) *Frontiers of Astrophysics*, Harvard University Press, Cambridge, MA, USA.

Cohen-Tanoudji, C. (1992) *Principles of Quantum Mechanics*, John Wiley and Sons, New York, NY, USA.

Dirac, P. A. M. (1982) *Principles of Quantum Mechanics*, 4th edn, Oxford University Press, Oxford, UK.

Griffiths, D. (1995) *Introduction to Quantum Mechanics*, Prentice Hall, Englewood Cliffs, NJ, USA.

Landau, L. D. and Lifschitz, E. M. (1977) *Quantum Mechanics*, 3rd edn, Pergamon Press, Oxford, UK.

Landau, R. H. (1996) *Quantum Mechanics II: A Second Course in Quantum Theory*, 2nd edn, John Wiley and Sons, New York, NY, USA.

Osterbrock, D. E. (1974) *Astrophysics of Gaseous Nebulae*, W. H. Freeman, San Francisco, CA, USA.

Shu, F. H. (1992) *The Physics of Astrophysics: Radiation*, University Science Books, Mill Valley, CA, USA.

Chapter 16

Molecular lines in astrophysics

The phase of the interstellar medium that is predominantly molecular contains many clues to the formation of stars and planets and may provide the link that connects molecular astrophysical gases to the formation of life. The fact that molecules are multi-nuclear and multi-electronic gives them a rich structure of energy levels. The complex structure is fortunately mediated by the existence of separate regimes of energy corresponding to the rotational, vibrational and electronic degrees of freedom. I will now discuss the energy level structure of molecules starting with a detailed discussion of the simplest systems, the diatomic molecules.

16.1 Diatomic molecules

16.1.1 Inter-nuclear potential

Given that atoms are characterized by the spatial scale length, a_0, we expect that two atoms will interact strongly when they are separated by a distance that is roughly a_0. Figure 16.1 shows that the inter-nuclear potential of two atoms has a minimum near $\Delta r \approx a_0$.

16.1.2 Electronic transitions

The energy scale of electronic transitions is given by the ionization potential

$$E_{el} \approx \frac{e^2}{a_0} = \hbar \omega$$

$$\Rightarrow \omega_{el} \rightarrow \text{visible and UV.}$$

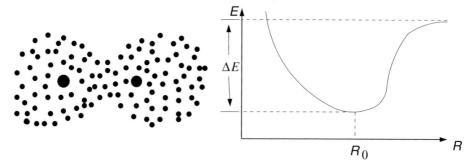

Fig. 16.1 The inter-nuclear potential of a diatomic molecule. The figure on the left shows the two nuclei surrounded by "electron clouds". The figure on the right shows a minimum in the potential for a separation near the Bohr radius.

16.1.3 Vibrational transitions

For a pair of nuclei, A and B, we can use a harmonic oscillator treatment

$$M\omega_{vib}^2 a_0^2 \approx E_{el}$$

$$\Rightarrow \omega_{vib} = \left(\frac{E_{el}}{Ma_0^2}\right)^{1/2} = \left(\frac{\hbar^2}{m_e Ma_0^4}\right)^{1/2}$$

so that

$$E_{vib} = \hbar\omega_{vib} \approx \frac{\hbar^2}{m_e^{1/2} M^{1/2} a_0^2} \approx \left(\frac{m_e}{M}\right)^{1/2} E_{el}.$$

We see that the energy range is reduced by the factor

$$\left(\frac{m_e}{M}\right)^{1/2}$$

relative to the electronic energy range. For $m_e/M \approx 10^{-3}$–$10^{-4} \rightarrow \omega_{vib} \rightarrow$ IR. The resulting transitions are in the infrared.

16.1.4 Rotational transitions

The rotation of the molecule represents a form of angular momentum which must also be quantized. The moment of inertia is defined as

$$I \approx Ma_0^2$$

so that the angular momentum of rotation becomes

$$L \approx I\omega_{rot} = Ma_0^2 \omega_{rot}.$$

Since the basic unit of L is $\hbar$ we have

$$Ma_0^2\omega_{rot} = \hbar$$

$$\Rightarrow E_{rot} = \hbar\omega_{rot} \approx \frac{\hbar^2}{Ma_0^2} \approx \left(\frac{m_e}{M}\right) E_{el}.$$

Typically, ω_{rot} is in the radio portion of the spectrum.

16.1.5 Summary

We see that the ratio of E_{rot} to E_{vib} to E_{el} has increments of $(M/m_e)^{1/2} \gtrsim 40$ so that the whole problem is amenable to perturbation analysis. In light of that we will now discuss the *Born–Oppenheimer Approximation*.

16.2   Diatomic molecules with two valence electrons

Consider a molecule with the indicated separations (Fig. 16.2). Since we are interested in the basic structure of the molecule let us neglect effects such as spin–orbit

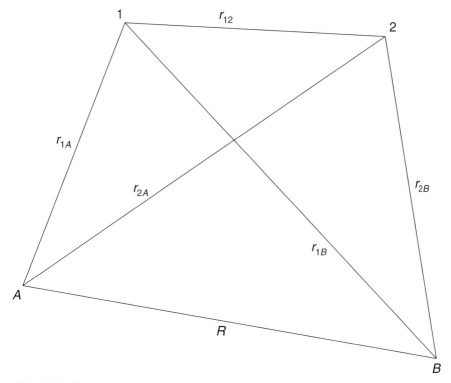

Fig. 16.2 The interaction associated with two valence electrons.

coupling and spin–spin interactions. The Hamiltonian now becomes

$$H = H_{AB} + H_{el} = \frac{-\hbar^2}{2M_A}\nabla_A^2 - \frac{\hbar^2}{2M_B}\nabla_B^2 - \frac{\hbar^2}{2m_e}\left(\nabla_1^2 + \nabla_2^2\right) + V \qquad (471)$$

where the A and B refer to the nuclei A and B and where V represents all the potentials

$$V = -\frac{Z_A e^2}{r_{1A}} - \frac{Z_B e^2}{r_{1B}} - \frac{Z_A e^2}{r_{2A}} - \frac{Z_B e^2}{r_{2B}} + \frac{Z_A Z_B e^2}{R} + \frac{e^2}{r_{12}}.$$

We must be careful to keep track of all the spatial variables and recognize that $\vec{\nabla}_A, \vec{\nabla}_B, \vec{\nabla}_1, \vec{\nabla}_2$ operate on $\vec{x}_A, \vec{x}_B, \vec{x}_1, \vec{x}_2$ respectively. We also note the relationships

$$R = |\vec{x}_A - \vec{x}_B|, \qquad r_{ij} = |\vec{x}_i - \vec{x}_j|.$$

16.2.1 The Born–Oppenheimer Approximation

Whenever we have used perturbation theory we have tried to separate wave functions into products of wave functions corresponding to the separate energy regimes. Since, $H_{AB} \approx (m_e/M)H_{el}$ we can try

$$\phi \approx \phi_{el}(\vec{x}_1, \vec{x}_2, R)\phi_{AB}(\vec{x}_A, \vec{x}_B). \qquad (472)$$

We can now set up a time-independent Schrödinger equation using (471) and (472) so that

$$\phi_{AB} H_{el}\phi_{el} + H_{AB}\phi_{el}\phi_{AB} = E\phi_{el}\phi_{AB}. \qquad (473)$$

If we now recall Leibnitz's rule

$$H_{AB}\phi_{el}\phi_{AB} = \phi_{el}H_{AB}\phi_{AB} - \frac{\hbar^2}{M_A}\vec{\nabla}_A\phi_{el} \cdot \vec{\nabla}_A\phi_{AB}$$

$$- \frac{\hbar^2}{M_B}\vec{\nabla}_B\phi_{el} \cdot \vec{\nabla}_B\phi_{AB} + \phi_{AB}H_{AB}\phi_{el}. \qquad (474)$$

The second and third terms can be combined to read

$$\hbar^2 \frac{\partial\phi_{el}}{\partial r}\hat{R} \cdot \left(\frac{1}{M_B}\vec{\nabla}_B - \frac{1}{M_A}\vec{\nabla}_A\right)\phi_{AB} \qquad (475)$$

because, $\vec{\nabla}_A\phi_{el} = \hat{R}(\partial\phi_{el}/\partial R) = -\vec{\nabla}_B\phi_{el} \quad [\vec{R} = \vec{x}_A - \vec{x}_B]$.

For a bound molecule the quantity (475) $\to 0$, on average. Furthermore, the fourth term in (474) is much smaller than the left-hand side of (473). Thus, the only important term is

$$H_{AB}\phi_{el}\phi_{AB} \approx \phi_{el}H_{AB}\phi_{AB}.$$

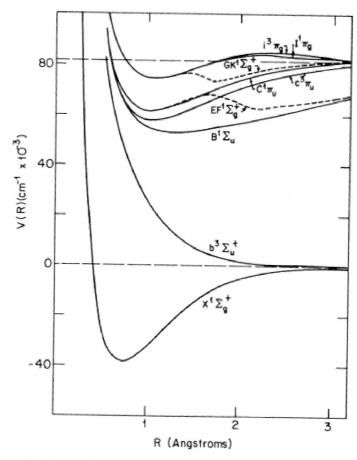

Fig. 16.3 The electronic energy levels of H_2. The labeling of the energy states is as follows. The ground state ($n = 1$) is labeled as X, while the excited states, $n = 2, 3, 4, \ldots$, are labeled $B, C, D, \ldots$, respectively. The corresponding unbound states are labeled $b, c, d \ldots$ The electronic orbital angular momentum is denoted by Λ and represents the momentum about the internuclear axis. Values are given by $\Lambda = \Sigma, \Pi, \Delta$ for 0, 1, 2, respectively. The left superscript represents the number of spin states (e.g. triplet, singlet). The right superscript refers to the symmetry of the electronic wave function upon reflection about any plane containing the two nuclei. The right subscript refers to the symmetry of the electronic wave function relative to the center of the molecule, which, in the case of H_2, is the center of symmetry of the electric field through which the electrons move. (From Shull, M. T. and Beckwith, S., 1982, ARA&A, **20**, pp. 163–190.) With permission, from the *Annual Review of Astronomy and Astrophysics*, Volume 20 © 1982 by Annual Reviews www.annualreviews.org.

Substituting into (473) yields

$$\frac{H_{el}\phi_{el}}{\phi_{el}} = -\frac{1}{\phi_{AB}} H_{AB}\phi_{AB} + E.$$

Note that the right-hand side now depends only on $\vec{x}_A$ and $\vec{x}_B$ while the left-hand side depends only on $\vec{x}_1$ and $\vec{x}_2$. The equation can be separated according to

$$H_{el}\phi_{el} = E_{el}(R)\phi_{el} \qquad (476)$$

$$H_{AB}\phi_{AB} + E_{el}(R)\phi_{AB} = E\phi_{AB}. \qquad (477)$$

Equation (476) describes the energy states of the electrons and can be solved using Coulomb and exchange integrals in a manner similar to that we used in the case of the helium atom. Figure 16.3 illustrates the first few electronic energy levels of the H_2.

Equation (477) describes the states associated with the motion of the nuclei (the vibrational and rotational degrees of freedom). We see that the Born–Oppenheimer Approximation has yielded the desired result of separating the Schrödinger equations. We now go on to show that (477) can be further separated according to translational, vibrational and rotational degrees of freedom.

16.3 Translational and internal degrees of freedom

To separate the internal degrees of freedom from bulk translational motions we need to consider the behavior of the molecule relative to its center of mass. We begin by introducing the center-of-mass (CM) coordinate

$$\vec{X}_{CM} \equiv \frac{M_A\vec{X}_A + M_B\vec{X}_B}{M_A + M_B}.$$

Defining the relative displacement, $\vec{R} = \vec{X}_A - \vec{X}_B$, the total mass, $M = M_A + M_B$ and the reduced mass $\mu = (M_A M_B)/(M_A + M_B)$ allows us to transform H_{AB} to CM coordinates such that

$$H_{AB} = -\frac{\hbar^2}{2M}\vec{\nabla}_{CM} - \frac{\hbar^2}{2\mu}\nabla_R^2$$

where, $\vec{\nabla}_{CM}$ and $\vec{\nabla}_R$ operate separately on $\vec{X}_{CM}$ and $\vec{R}$ respectively. We therefore look for eigenfunctions of the form

$$\phi_{AB} = \phi_{trans}(\vec{X}_{CM})\phi_{int}(\vec{R}).$$

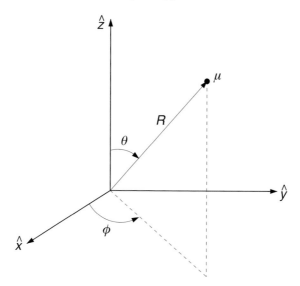

Fig. 16.4 The geometry used to define the angular momentum.

Substitution into (477) yields

$$\frac{1}{\phi_{trans}}\left(-\frac{\hbar^2}{2M}\nabla_{CM}^2\phi_{trans}\right) = \frac{1}{\phi_{int}}\left(\frac{\hbar^2}{2\mu}\nabla_R^2\phi_{int}\right) - E_{el}(R) + E. \quad (478)$$

Since ∇_{CM}^2 and ∇_R^2 operate independently, both sides must equal a constant

$$\Rightarrow -\frac{\hbar^2}{2M}\nabla_{CM}^2\phi_{trans} = E_{trans}\phi_{trans} \quad (479)$$

$$-\frac{\hbar^2}{2\mu}\nabla_R^2\phi_{int} + E_{el}(R)\phi_{int} = E_{int}\phi_{int} \quad (480)$$

where $E_{int} = E - E_{trans}$.

We see that we have achieved our goal of isolating the internal degrees of freedom. All that is left now is to separate them into vibrational and rotational degrees of freedom.

16.3.1 Vibrations and rotations

To separate (480) into vibrational and rotational components we begin by introducing spherical coordinates for R (Fig. 16.4)

$$\nabla_R^2 = \frac{1}{R^2}\left[\frac{\partial}{\partial R}\left(R^2\frac{\partial}{\partial R}\right) - \frac{L^2}{\hbar^2}\right] \quad (481)$$

where $\vec{L} = \vec{R} \times (-i\hbar\vec{\nabla}_R)$, as usual.

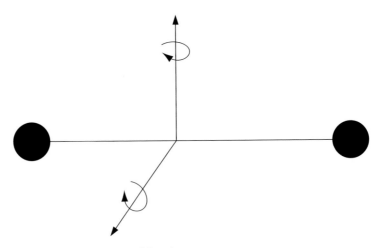

Fig. 16.5 Rotational degrees of freedom.

E_{el} depends only on $|R|$ so that (480) can be separated into radial and angular parts so that

$$\phi_{int}(R, \theta, \phi) = \frac{1}{R} Z_{vib}(R) Y_{Jm}(\theta, \phi)$$

where J, m are analogous to ℓ, m. Thus

$$L^2 Y_{Jm} = J(J+1)\hbar^2 Y_{Jm}$$

$$L_Z Y_{Jm} = m\hbar Y_{Jm}$$

also in analogy to the angular momentum operator in atomic physics.

If we now substitute (481) into (480) using these relations we get

$$-\frac{\hbar^2}{2\mu} \frac{d^2 Z_{vib}}{dR^2} + E_{el}(R) Z_{vib} = \left[E_{int} - \frac{J(J+1)\hbar^2}{2\mu R^2} \right] Z_{vib} \qquad (482)$$

which is Schrödinger's equation for anharmonic oscillators in which Y_{Jm} represents the two independent degrees of freedom for directional orientation of the diatomic molecules, that is two rotations (Fig. 16.5).

Now let us consider the radial degree of freedom, the vibrations, and we will have separated out the three regimes of motion.

16.3.2 Vibrations – harmonic oscillator approximation

Let us assume small excursions relative to the mean separation, R_0, where $E'_{el}(R_0) = 0$, using a Taylor series expansion so that

$$E_{el}(R) = E_{el}(R_0) + \frac{E''_{el}(R_0)}{2}(R - R_0)^2 + \cdots$$

If we now define $\mu\omega_0^2/2 = E_{el}''(R_0)/2$ and $x = R - R_0$ we can write the radial part of (482) as

$$-\frac{\hbar^2}{2\mu}\frac{d^2 Z_{vib}}{dx^2} + \frac{\mu}{2}\omega_0^2 x^2 Z_{vib} = E_{vib}Z_{vib}. \qquad (483)$$

In (483) we have defined

$$E_{vib} = E_{int} - E_{el}(R_0) - \frac{J(J+1)\hbar^2}{2\mu R_0^2}$$

that is $E_{int} = E_{el}(R_0) + E_{vib} + E_{rot}$.

Thus, we identify

$$E_{rot} = \frac{J(J+1)\hbar^2}{2\mu R_0^2} = J(J+1)B \qquad J = 0, 1, 2, \ldots$$

where B is the *rotational constant* of the molecule.

Equation (483) is the Schrödinger equation for a harmonic oscillator. The eigenvalues are therefore

$$E_{vib} = \left(v + \frac{1}{2}\right)\hbar\omega_0 \qquad v = 0, 1, 2, \ldots$$

Now that we have defined the energy level structure of diatomic molecules let us consider transition rates in order to determine what transitions are actually possible on these energy ladders.

16.4 Dipole transition probability

In the dipole approximation we can define a perturbation Hamiltonian as follows

$$H_d = -\vec{E}\cdot\vec{d}$$

where $\vec{d}$ is the dipole moment and $\vec{E}$ is the electric field of the radiation field. The dipole moment of a diatomic molecule is defined as

$$\vec{d} \equiv \vec{d}_{el} + \vec{d}_{nuc} = \sum_{a=1}^{N}\vec{d}_a + Z_A e\vec{X}_A + Z_B e\vec{X}_B$$

where $\vec{d}_a \equiv -e\vec{X}_a$ is the dipole moment of the ath electron relative to the center of mass of the molecule. The dipole transition matrix element can be defined in the usual way so that

$$\langle\Psi_f|H_d|\Psi_i\rangle = -\langle\Phi_f|\vec{E}_\omega\cdot(\vec{d}_{el} + \vec{d}_{nuc})|\Phi_i\rangle \qquad (484)$$

where

$$\Phi = \frac{1}{R} Z_{vib}(R) Y_{Jm}(\theta, \phi) \phi_{trans} \phi_{el} \Sigma$$

where Σ is the electron spin wave function.

16.4.1 Pure rotational spectra

Let us consider the lowest energy transitions first. These are the transitions in which the initial and final vibrational and electronic states are the same. Equation (484) then becomes

$$\langle \Psi_f | H_d | \Psi_i \rangle = \int_0^{2\pi} d\phi \int_0^{\pi} Y_{Jm}^*(\theta, \phi) \langle Z_{vib} | H_D | Z_{vib} \rangle Y_{J'm'}(\theta, \phi) \sin\theta \, d\theta \quad (485)$$

where

$$\langle Z_{vib} | H_D | Z_{vib} \rangle \equiv -\int_0^\infty \vec{E}_\omega \cdot \vec{d}(R, \theta, \phi) |Z_{vib}(R)|^2 \, dR$$

and where

$$\vec{D}(R, \theta, \phi) \equiv \int (\vec{d}_{el} + \vec{d}_{nuc}) |\phi_{el}(\vec{X}_1, \vec{X}_2, \ldots, \vec{X}_N; R)|^2 \, d^3 X_1 \, d^3 X_2 \ldots d^3 X_N.$$

The symbol $\vec{D}$ is commonly referred to as the permanent dipole moment. Thus, for a transition to have a nonvanishing probability the permanent dipole moment must not be zero.

For a diatomic molecule, the rather specific geometry allows a simple expression for the nuclear dipole moment

$$\vec{D}_{nuc} = \vec{d}_{nuc} = \frac{e}{M}(Z_A M_B - Z_B M_A)\vec{R} = D(R)\frac{\vec{R}}{R}.$$

Consider now the case where

$$\vec{E}_\omega \cdot \vec{D} = |\vec{E}_\omega| D(R) \cos\theta \approx |\vec{E}_\omega| D_0 \cos\theta$$

where θ is the angle between the electric vector and the z-axis and where we have made the approximation that $D(R) = D(R_0) \approx D_0$ which is valid when the molecule is in the vibrational and electronic ground state.

Equation (485) now becomes

$$\langle \Psi_f | H_d | \Psi_i \rangle = -|\vec{E}_\omega| D_0 \int_0^{2\pi} d\phi \int_0^{\pi} \cos\theta Y_{Jm}^*(\theta, \phi) Y_{J'm'} \sin\theta \, d\theta.$$

Table 16.1. *Molecular transition rates*

Molecule	Transition	Rest frequency (MHz)	A (sec^{-1})
$^{12}C^{16}O$	$J = 1 \rightarrow 0$	115271.2	6×10^{-8}
	$J = 2 \rightarrow 1$	230542.4	$-$
$^{13}C^{16}O$	$J = 1 \rightarrow 0$	110201.4	$-$
$^{1}H_2^{12}C^{16}O$	$J_{K^-K^+} = 1_{10} \rightarrow 1_{11}$	4829.66	3×10^{-8}
	$2_{11} \rightarrow 2_{12}$	14488.65	3×10^{-7}
	$2_{02} \rightarrow 1_{01}$	145602.97	$-$
	$2_{12} \rightarrow 1_{11}$	140839.53	5×10^{-5}
	$2_{11} \rightarrow 1_{10}$	150498.36	$-$
$^{16}O^{1}H$	$J = 2/3, F = 1 \rightarrow 2$	1612.231	1.3×10^{-11}
	$1 \rightarrow 1$	1665.402	7.7×10^{-11}
	$2 \rightarrow 2$	1667.359	7.7×10^{-11}
	$2 \rightarrow 1$	1720.530	9×10^{-12}

The above allows us to construct the transition dipole moment

$$|\vec{d}_{J+1,J}|^2 \equiv \sum_{m=-J}^{J} D_0^2 \left| \int Y_{Jm}^* \cos\theta\, Y_{J+1,m'}\, d\Omega \right|^2 = \frac{J+1}{2J+3} D_0^2. \qquad (486)$$

With the transition dipole moment defined we can now use the standard formula for A to get the transition probability

$$A_{J+1,J} = \frac{4\omega^3}{3\hbar c^3} |\vec{D}_{if}|^2 = \frac{4(J+1)\omega^3}{3(2J+3)\hbar c^3} D_0^2. \qquad (487)$$

Equations (486) and (487) allow us to calculate the dipole transition rate between adjacent rotational levels of molecules. Some rotational transition rates are listed in Table 16.1.

The selection rule for rotational transitions is $\Delta J = \pm 1$.

16.5 Transitions between vibrational levels

Similar considerations apply for vibrational levels. We proceed by replacing the vibrational integral in (485) with $\langle v'|H_D|v\rangle$. The selection rule is given by, $\Delta v = \pm 1$.

In general we have to allow for a change in rotational as well as vibrational levels. The energy level associated with the transition is therefore given by

$$E_{vJ} = \left(v + \frac{1}{2}\right)\hbar\omega_0 + J(J+1)B.$$

For the transition $v' \rightarrow v$ such that $\Delta v = \pm 1$, J can change by either $+1$ or -1. Thus, the frequency of the emitted photon has two sets of possibilities

$$\omega = \omega_0 + 2(J+1)B/\hbar \qquad J' = J + 1 \qquad \rightarrow R \text{ branch}$$
$$\omega = \omega_0 - 2JB/\hbar \qquad J' = J - 1 \qquad \rightarrow P \text{ branch}.$$

Paradoxically the natural vibrational frequency, ω_0, is not possible because it requires $\Delta J = 0$ and it would violate the dipole selection rules. Instead there is a band of rotational transitions associated with each ω_0. The band is made up of the P and R branches.

The dipole selection rules for rotational–vibrational transitions are $\Delta v = \pm 1$, $\Delta J = \pm 1$.

16.6 Transitions between electronic levels

The retention of ϕ_{el} in (484) and (485) allows the calculation of transition rates between levels having different electronic, vibrational *and* rotational states.

Electronic transitions do not require the molecule to have a dipole moment because the electronic dipole moment is governed by the charge distributions associated with the electronic levels as in the atomic case.

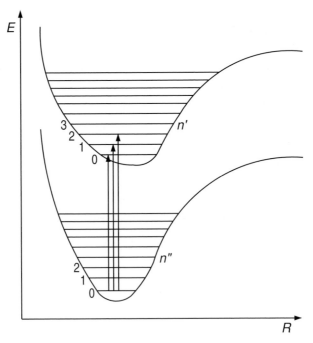

Fig. 16.6 Electronic transitions. Note the splittings that arise from the available vibrational levels.

Consequently, the ΔJ can be 0 (so long as $\Delta \Lambda \neq 0$) and there is no restriction on Δv. The dipole selection rules are therefore $\Delta J = 0, \pm 1$, $\Delta m = 0, \pm 1$, $\Delta \Lambda = 0, \pm 1$ and $\Delta S = 0$.

The $\Delta J = 0$ possibility introduces another branch, the Q branch which complements the P and R branches and adds to the wealth of the band structure in molecular spectra. Thus, for every electronic transition $n' \to n$, there is fine structure associated with the unrestricted vibrational transitions (see Fig. 16.6).

Each vibrational transition, in turn, consists of a band of rotational transitions adding even finer structure. The final result is a rich structure of lines associated with each electronic transition.

We now examine two specific molecules to illustrate some of these concepts and to describe how their spectra are used as diagnostics of the interstellar medium.

16.7 The H₂ molecule

We now examine the simplest possible molecule, H_2. According to (483), the $J = 1 \to J = 0$ rotational energy transition should have an energy gap of $2B$, where the rotational constant

$$B = \frac{\hbar^2}{2\mu R_0^2} = 10^{-14}$$

for H_2. Thus, $\hbar\omega = 2 \times 10^{-14}$ ergs $= 0.014$ eV so that the emitted frequency is $\nu \approx 1.6 \times 10^{12}$ Hz ($\lambda \approx 200\ \mu$m).

There are no radio-frequency lines. The expected transitions fall into the far IR window of the spectrum. Do we see these transitions? According to (484), the dipole transition rate depends on the permanent dipole moment of the molecule. For H_2 this dipole moment is 0 because of the obvious symmetry in the nuclear charge distribution and the consequent symmetry of the charge to rotations. Furthermore, the H_2 molecule is not detectable at radio frequencies, which is unfortunate because radio frequencies are the only ones immune to galactic extinction (attenuation by interstellar dust). Thus, once in the molecular form, hydrogen becomes very difficult to detect. We must turn to vibrational transitions. Recall from (483) that

$$E_{vib} = \left(v + \frac{1}{2}\right)\hbar\omega_0$$

where ω_0 is the natural or fundamental vibrational frequency of the molecule which can be measured in the laboratory. For H_2, therefore, $\omega_0 = 3.3 \times 10^{14}$ radians per second and $\nu = 5 \times 10^{13}$ Hz which is in the near IR part of the spectrum (wavelength of 6 micrometers). Figure 16.7 illustrates the vibrational levels of H_2. These are subject to the selection rule, $\Delta v = \pm 1$. Vibrational transitions can involve changes

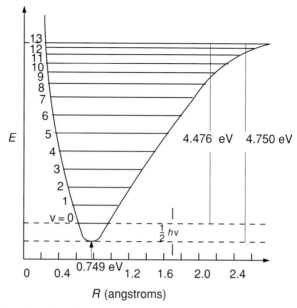

Fig. 16.7 Vibrational levels of H_2.

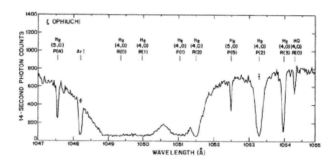

Fig. 16.8 Observed spectrum of H_2 in the UV portion of the spectrum near 1000 Å. (From Spitzer, L. and Jenkins, J. B., 1975, ARA&A, 13, 133–164.) With permission, from the *Annual Review of Astronomy and Astrophysics*, Volume 20 © 1982 by Annual Reviews www.annualreviews.org.

in rotational levels so that an overall modulation of the vibrational transition leads to multiple spectral lines.

The electronic transitions have even greater energy gaps. Figure 16.8 shows the UV spectrum of the electronic transition $B^1\Sigma_u^+ \rightarrow X^1\Sigma_g^+$. Note the bands in the UV caused by the blending of lines from the P and R branches. Figure 16.9 illustrates schematically some of the lines visible in Fig. 16.8. The spectrum is obtained by observing UV absorption in the direction of luminous blue stars. These kinds of observations have revealed the presence of a diffuse interstellar

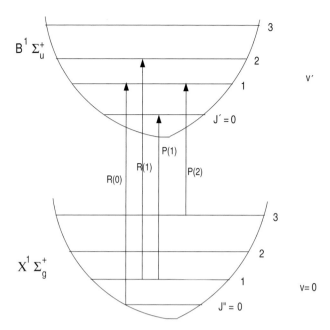

Fig. 16.9 Examples of rotational transitions associated with a single electronic transition. The labeled transitions correspond to rotational fine structure lines seen in Fig. 16.8.

molecular gas. Observations of IR lines from vibrational transitions have confirmed the widespread existence of H_2. Analysis of the H_2 spectra has shown that the diffuse molecular hydrogen has a density of $n = 10$–10^3 cm^{-3} and a temperature $T = 10^2$–10^3 K.

The absence of radio lines in the spectra of H_2 precludes probing the distribution of H_2 in the centers of molecular clouds where dust extinction is severe. Consequently, studies of H_2 tell us relatively little about the interiors of molecular clouds. We must turn to a molecule that has rotational radio-frequency transitions.

16.8 The CO molecule

The CO molecule consists of a ^{12}C and a ^{16}O nucleus along with the associated electrons. Its reduced mass is therefore $(12 \times 16/28) = 6.9$ nucleons so that its B is 13.8 times smaller than that of H_2. The resulting $J = 1 \rightarrow 0$ transition frequency is 1.15×10^{11} Hz or a wavelength of 2.6 mm. This transition does fall into the radio portion of the EM spectrum. Furthermore, since the molecule has a permanent dipole moment there is a finite dipole transition rate so that the transition can be actually observed.

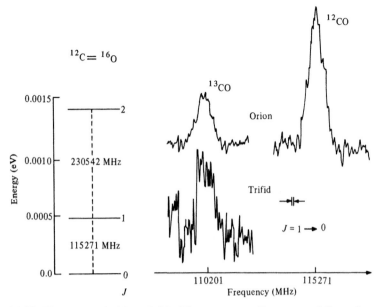

Fig. 16.10 The energy ladder of CO. The corresponding spectral lines from the Orion and Trifid nebulas are also shown. Reprinted by permission of the publisher from FRONTIERS OF ASTROPHYSICS, edited by Eugene Avrett, Cambridge, Mass,: Harvard University Press, Copyright © 1976 by the Smithsonian Institution Astrophysical Observatory.

For CO, $D = 0.112 \times 10^{-18}$ esu cm so that the transition rate according to (217) is $A_{10} = 6 \times 10^{-8}$. The value of A is low only because of the low frequency of the transition. It is *not* a forbidden transition, it satisfies the dipole selection rule, $\Delta J = \pm 1$. Figure 16.10 shows observations of this transition from the Orion and Trifid nebulas. Observations of the line strength provide information about the temperature and density of the molecule's environment since the rotational levels are populated by collisions with neighboring molecules (mainly H_2 as it turns out). Figure 16.11 shows the dependence of the line strength on the temperature and density. Observations of the spectral lines of the isotope $^{13}C\ ^{16}O$ yield additional estimates of the ambient density. Thus, CO observations are excellent probes of molecular clouds.

Vibrational–rotational transitions of CO illustrate nicely the band structures associated with the P and R branches as shown in Fig. 16.12 for the $v = 1 \to 0$ transition. Transitions associated with J as high as 23 are noticeable. These lines are found near a wavelength of 5 μm. Modern observations allow maps of emission to be made, as shown in Fig. 16.13.

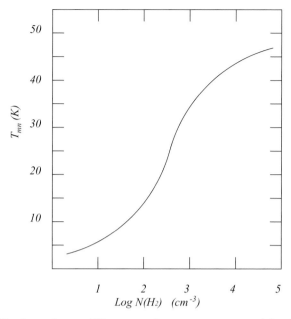

Fig. 16.11 The dependence of line strength on temperature and density.

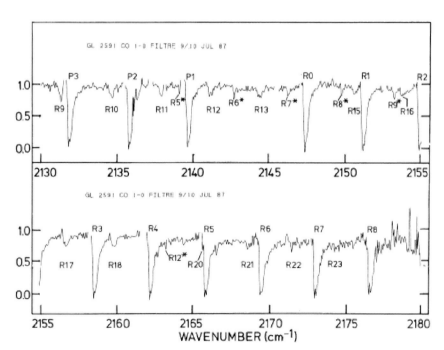

Fig. 16.12 Observed vibrational–rotational transitions of CO. (From: Mitchell, George F.; Curry, Charles; Maillard, Jean-Pierre; Allen, Mark, 1989, ApJ, 341, pp. 1020–1034.)

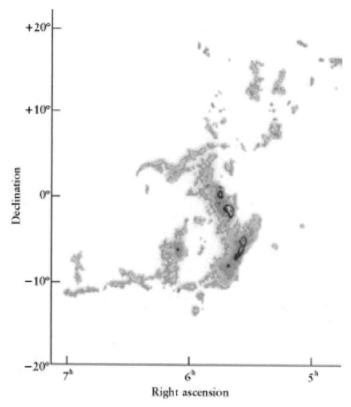

Fig. 16.13 Maps of CO emission from the Orion nebula. The grey-scale representation shows the distribution of line intensities for the $J = 1 \rightarrow 0$ transition of CO. (From: http://rst.gsfc.nasa.gov/sect20/A5.html.)

16.9   Other molecules

There is a great diversity of molecular lines arising from a plethora of molecules which, when taken together, yield a complete picture of the conditions inside molecular clouds. Fig. 16.14 shows spectral features arising from H_2CO. The study of molecules such as CO, OH and H_2CO and others has led to a detailed understanding of the environments in which stars are born. Such studies are crucial to an understanding of how stars (like our Sun) and planets (like our solar system) are born. In the 1980s and 1990s the conventional view that stars form from a symmetric collapse of a spherical cloud has been turned on its head. We now know that molecular clouds form flat disks inside which the forming stars drive massive outflows of molecular gas out of the collapse regions. These YSOs (young stellar objects) are beacons of star formation in much the same way that radio continuum jets and lobes are indicators of active galactic nuclei and quasars.

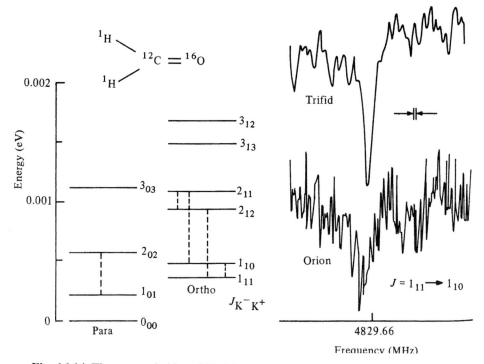

Fig. 16.14 The energy ladder of H_2CO. Also shown are the spectral lines from the Orion and Trifid nebulas. Reprinted by permission of the publisher from FRONTIERS OF ASTROPHYSICS, edited by Eugene Avrett, Cambridge, Mass,: Harvard University Press, Copyright © 1976 by the Smithsonian Institution Astrophysical Observatory.

16.10 Life in the Universe

I close out the book by briefly discussing the question – is there other life in the Universe? I discuss this from two points of view. First, we ask whether formation of life is a universal process. Does it occur as a natural process or is it a fluke? Secondly, I explore the optimal way to attempt communication with other potential civilizations.

16.10.1 The building blocks of life

The fundamental basis of life on Earth is the complex carbon-based organic molecules that are found in all terrestrial life forms. The interactions of these molecules are optimized by the presence of water. Molecular clouds are so dense that complex molecules can form and be protected from dissociation. The identifiable features include absorption bands (like the P, Q and R branches I previously discussed) of silicate grains (the same as that found in minerals on Earth) and water ice (H_2O). The formation of water, at least, appears to be a ubiquitous process.

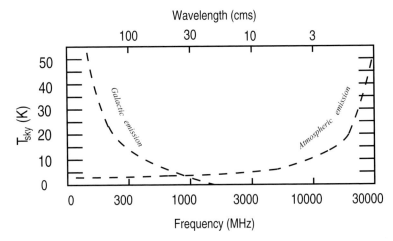

Fig. 16.15 The radio spectrum of emission from the Galaxy and the Earth's atmosphere. Note the emission minimum near 1 GHz. The minimum represents a natural observing window optimized for the HI and OH lines.

What is even more astonishing is the evidence for organic molecules, which are required to explain the details of the water bands seen in these spectra.

The presence of water and organic molecules in molecular clouds does not indicate that these are then incorporated into stars and planets (they are not, because they are destroyed near the forming star) but rather that these molecules are a natural by-product of universal processes. Thus, if such molecules can be formed in molecular clouds they can certainly form on the surfaces of newly formed planets. This does not of course prove that life itself is a universal process but it is evidence in that direction.

16.10.2 *Communicating with other civilizations*

We are technologically capable of sending and receiving communication signals to a distance of a few tens of parsecs. Such distances encompass many Sun-like stars. If there are technologically advanced civilizations among them we should be able to communicate with them. The question is – what is the best way? The radio part of the spectrum is the obvious place to begin because radio waves are not attenuated by atmospheres and other gases and dust between us and the source. However, the radio spectrum itself suffers from two major sources of noise. Radio emission from the Earth's atmosphere is strong at wavelengths shorter than 3 cm. Radio emission from the Milky Way is strong at wavelengths longer than 30 cm. Thus, the obvious window to which we can narrow our search is between 3 and 30 cm (see Fig. 16.15).

It turns out that the neutral hydrogen line (HI) lies at 21 cm while strong lines of OH can be found at 18 cm. Perhaps these are natural wavelengths at which to search for artificial signals. Alternatively, one might attempt to search for signals between these wavelengths since this region is relatively free of natural signals. Multi-channel searches of this region for artificial signals are currently being carried out. So far there are no results but we have only just begun.

16.11   Further reading

Atkins, P. W. and Friedman, R. S. (1999) *Molecular Quantum Mechanics*, Oxford University Press, Oxford, UK.

Avrett, E. H. (Ed) (1976) *Frontiers of Astrophysics*, Harvard University Press, Cambridge, MA, USA.

Goldsmith, D. and Owen, T. (2001) *The Search for Life in the Universe*, University Science Books, Mill Valley, CA, USA.

Shu, F. H. (1992) *The Physics of Astrophysics: Radiation*, University Science Books, Mill Valley, CA, USA.

Verschuur, G. A. and Kellerman, K. I. (1988) *Galactic and Extragalactic Radio Astronomy*, Springer-Verlag, Berlin, Germany.

Index